SpringerBriefs in Energy

More information about this series at http://www.springer.com/series/8903

Taofeek Orekan • Peng Zhang

Underwater Wireless Power Transfer

Smart Ocean Energy Converters

 Springer

Taofeek Orekan
Electrical and Computer Engineering
University of Connecticut
Storrs, CT, USA

Peng Zhang
Electrical and Computer Engineering
University of Connecticut
Storrs, CT, USA

ISSN 2191-5520   ISSN 2191-5539   (electronic)
SpringerBriefs in Energy
ISBN 978-3-030-02561-8   ISBN 978-3-030-02562-5   (eBook)
https://doi.org/10.1007/978-3-030-02562-5

Library of Congress Control Number: 2018960758

This Springer imprint is published by the registered company Springer Nature Switzerland AG
The registered company address is: Gewerbestrasse 11, 6330 Cham, Switzerland

*This book is dedicated to my fiancée,
Dr. Lingyu Ren, for her devotion, kindness,
and endless support. Also, I dedicate this
to my father and mother: Hakeem &
Iyabo Orekan.*

To Haizhen, William, Henry, and Benjamin

Contents

Chapter 1
Overview of the Smart Ocean Energy Converter

1.1 Introduction

There is a need in the foreseeable future for ocean system electrification, renewable ocean power sources, and ocean energy network, which could help speed up the growth and deployment of marine renewable energy and ways of exploring and understanding the ocean. Some attributes of ocean energy such as its high-energy density that makes it attractive as a grid-connected energy source could also make it attractive as an isolated remote ocean energy source to provide power solutions for a sustainable development in the ocean space. It is advantageous to rapidly developing distributed ocean energy applications, such as underwater sensing networks, ocean sensors and monitoring technology, marine automated network buoys, and deep-ocean and tsunami buoys. In particular, it can supply power to autonomous underwater vehicles (AUVs), whose operating lifetimes are limited by their battery power.

At their most basic level, AUVs are plainly computer-controlled vehicles operating undersea. They are attractive option for ocean-based research, military purposes including underwater mine hunting, and increasingly popular in offshore oil industry to identify underwater cables and pipes that need repairs. As the AUVs become more sophisticated, one of the critical requirements to be able to achieve their full capability is to supply sufficient energy (electrical power) to operate. Although the requirements for improved power distribution, management, and reduced weight of batteries in AUV are currently being researched, most designs favor that of rechargeable batteries as the main power source, which may be swapped out and replaced by another fully charged battery. However, the high maintenance and replacement costs of batteries cannot offer sustainable operation [1]; the already burdensome tasks of carrying, operating, and maintaining multiple batteries will only worsen the burden. In addition, the daily operating cost of a ship for AUV operations ranges from $35k to $40k, without guarantee of operator and ships in bad weather. In ocean-based research, many of the important data and samples are

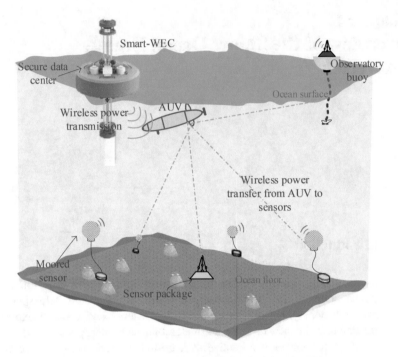

Fig. 1.1 Smart-WEC transferring power wirelessly to distributed ocean devices

usually collected during harsh weather. Likewise in military operations, AUV has to operate mostly at odd working hours.

Hence, it is necessary to rethink the battery burden in AUV and other distributed ocean systems. Power harvesting from ocean energy is a natural option to replace it. This book introduces a smart wave energy converter (Smart-WEC), a new type of power harvester that extracts energy from the motion of the ocean waves. It has the potential to resolve the long-standing challenge of providing sustainable power to distributed ocean applications, while addressing the limitations of conventional ocean technology to allow a more robust, smart, and reliable ocean energy systems. Most importantly, the Smart-WEC has the capability of making wireless connections with distributed ocean system such as AUV, through a novel underwater wireless power transfer system, as shown in Fig. 1.1. As a result, this technology could make a profound impact on the development of AUVs by reducing their battery burden. In an effort to assure the energy resilience and maximize power absorption, this book also discusses and explores new control theories, such as maximum power efficiency tracking (MPET) control, maximum power extraction control (MPEC) for Smart-WEC, and maximum life cycle tracking (MLCT) for tidal generation.

The rest of this chapter provides a brief literature review of ocean energy technology and an overview of underwater wireless power transfer. A number of significant prior works and challenges in the field of ocean energy and inductive wireless power transfer are summarized.

1.2 Literature

The world's interest in the development of the renewable sources of energy has increased rapidly in the past two decades. Humans are directing their attention to the ocean, which covers approximately 71% of the earth surface. About one-tenth of energy like tidal energy, wave energy, ocean thermal energy, and offshore wind energy that can be harnessed from the ocean is equivalent to five times of the world's energy demand [2]. And there has been recent evidence that 50% of the world's electricity consumption can be covered by wave energy alone [3]. In addition, ocean energy has the potential to make significant contribution to future alternative energy to more than 44% of the world's population living within 150 km of the sea [4].

Harnessing energy from the harsh ocean environment does not come, in fact, without challenges. Despite the enormous potential, however, many ocean harvesters such as wave energy converters and tidal turbines are still in developing stages—there are only a limited number of technologies that made their way to commercialization stage or connected to a main power grid. The major challenges are pertaining to the reliability and survivability of these technologies as well as a typical problem with renewable energy systems, intermittent of the power output. As a matter of fact, there have been critical failures of WEC and tidal turbine in recent years [5], most of which resulted in hindered development and in many cases shutdown of wave energy companies. Furthermore, traditional grids were not designed to cope with fluctuating electricity supply, which can create negative impact on the grid operation and stability [6–8].

1.2.1 Wave Energy

Seldom exploited energy source is the ocean wave which is the most prominent form of ocean energy, mainly because of the often conspicuous but destructive effects of the waves, which are generated by wind action, therefore making them an indirect form of solar energy.

1.2.1.1 Physics of Ocean Waves

Wind-produced waves have the most energy concentration than other types of waves observed in the ocean as illustrated in Fig. 1.2. These types of waves are created as winds blow across the oceans, the surface exerting the gravitational force on the bottom layer of the wind. This, in turn, exerts the pull until it reaches the

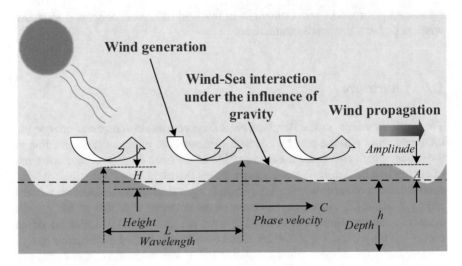

Fig. 1.2 Characteristics of a wave

topmost layer. Once made, wind waves can travel many kilometers with little energy depletion. Closer the coastline the wave energy force diminishes due to interaction with the seabed.

Ocean wave encompasses two forms of energy: the potential energy of elevated water particles and the kinetic energy of the water particles that flows in a circular path. The average kinetic energy in a linear wave is almost equal to potential energy. Figure 1.2 shows the characteristics of a wave. The power in the wave is proportional to the period of the motion and the square of the amplitude. Energy fluxes are usually between 50 and 70 kW/m width of oncoming wave. Thus, the long period (8–12 s) and the large amplitude (2–4 m) waves have interest for power generation.

1.2.1.2 Benefits

There are a variety of benefits by harvesting wave energy for electricity generation. Even though it is currently not economically competitive with other energy technologies, due to the high-power density of ocean waves, it has the potential to be competitive once the technology is matured. Below are some of the unique features of the ocean waves.

- *Predictability*: Ocean waves are consistent and have good forecast ability. Because waves propagate across the ocean, their time of arrival at the wave power plant may be predicted better than wind. With measurement buoys hundreds of miles away offshore, 10 h or longer forecasting can be achieved; power producers can use this information to calculate the amount of power and energy that it can produce and develop dispatched plans to meet electrical demands. With more advanced techniques such as satellite imagery, days ahead forecast could be possible.

- *High Availability and Power Density*: The typical power density of ocean waves is about typical 30 kw/m. This suggests wave energy has the capacity to become the lowest cost renewable energy resource. The power density of wave energy largely surpasses that of wind energy, as seawater is about 800 times denser than air. This dramatically increases the amount of energy available in wave energy.
- *Location*: Different wave energy conversions will be utilized at various locations to take advantage of the variability of ocean energy resource.

1.2.1.3 Wave Energy Technologies

Unlike other renewable energy technologies, there are several categories of wave energy converters. The four typical types of wave energy technology developed to capture the mechanical energy within waves and transform the captured energy into electricity are described in [9, 10]. An extensive literature review of the many proposed WEC designs is provided in [11]. Among the most popular design categories are the following: oscillating water columns, point absorbers, attenuators, and overtoppings. The most researched WEC device is the point absorber, which includes such models as the Wavestar, OPT Powerbuoy, and Wavebob [12]. Currently, a new type of WEC (variable geometry), inspired by wind energy, was developed in National Renewable Energy Laboratory (NREL). This type of WEC with horizontal flaps can modify its shape to change how it interacts with waves [13].

Transforming wave energy into electricity typically entails several stages. In most devices, the first conversion stage involves the conversion of wave energy into kinetic or potential energy. Then, a power take-off (PTO) mechanism must convert this kinetic or potential energy into electrical energy. In many cases, power electronic converters can be used to rectify or invert the electrical energy [14].

Seemingly similar to wind turbines, there are more multifarious of wave energy technologies than wind turbine. However, still no consensus is reached on the most appropriate WEC for resource exploitation. WEC development, vigorous testing, evaluation procedures, and assessments of the device performance are important issues. Numerical analysis or a physical modeling can be used to perform such assessments. An ideal WEC is a good wave maker operating in all six degrees of freedom allowing for almost 100% extraction of the wave energy coming toward it. In other words, there are incoming waves in front of the WEC and still water on the backside.

1.2.1.4 Power Take-Off Mechanism

The point absorber is a type of converter that is either floating or mounted on the seabed and absorbs energy in all directions through its movement at or near the

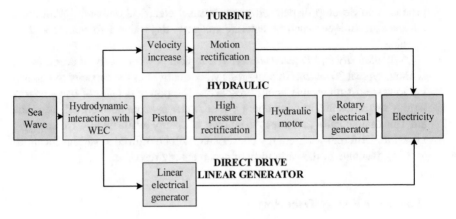

Fig. 1.3 Block diagram of generators used WEC technologies

surface of the sea. As seen in Fig. 1.3, the hydraulic power take-off (PTO) and linear generator are the two main types of generator used in the point absorber WEC. The linear generator can be directly linked to the system without gear box to generate power. Its advantage over conventional high-speed rotating generators or hydraulics is that the movement is directly converted into electricity and no secondary energy conversion is necessary [15].

1.2.1.5   Maximum Power Extraction

In WEC design, the incoming incident waves can be represented as a linear monochromatic waves, i.e., adjustable (frequency and amplitude) sinusoidal, and as nonlinear irregular waves. Furthermore, for the device to withstand the high loads and harsh environment of open water—what researchers call "slamming events"—designs and hydrodynamic assessments of the device performance are critically important [13]. Typically, hydrodynamics of a WEC is calculated with its linear coefficients, and evaluations of its performance are often based on the sinusoidal assumption of incident waves [16] behavior. Some researchers have suggested that for WEC operating in the power production mode, using linear wave theory can cover a large percentage (about 80%) of the wave resource in operational condition. However, WECs being device-dependent, this method is not suitable for verifying the actual performance of the device, and, in reality, ocean waves are highly irregular. Such method could only lead to incorrect predictions about power absorption and power production under partial nonlinear waves. Moreover, this renders control strategy (survival mode) implemented under such assumption unreliable, in the case of extreme waves.

Both theoretical and practical studies have been analyzed to maximize WEC systems' energy absorption. One such study proposed the use of a latching control method, which forcibly locks the WEC's motion when the velocity is close to zero and then releases it after a certain duration or "latching time" [17, 18]. This keeps the velocity in phase with the excitation force so that the maximum amount of power can be extracted. Although the use of a latching control method would theoretically allow for maximal energy conversion, its practicality has been questioned in [19] due to the additional mechanical configuration required to hold the WEC.

1.2.1.6 Challenges

Although research and development of wave energy technologies have progressed to commercial construction for meaningful power generation, it is important to appreciate the many difficulties hindering the rapid development of these technologies. Summarized below are some challenges facing wave power:

- It is extremely difficult to couple irregular wave motion, where the wave frequencies are typically ≈ 0.1 Hz, to power take-offs or electrical generators requiring almost 600 times greater frequency.
- The irregularity of the wave direction, phase, and amplitude makes it quite challenging designing technology that can extract power at high efficiency. In addition, the estimation of the wave energy potential (dependent on two wave parameters: wave height and wave period) is not as straightforward as in wind power.
- More attractive waves (peak power) are located in deep-water waves from open-sea swells produced from long fetches of prevailing winds. However, due to the difficulties in maintenance, fixing mooring system in deep sea, and transmitting electrical power to onshore, most WEC devices are located nearshore.
- To survive hurricanes and other extreme weather conditions, the WEC device structure has to withstand a hundred times of its power intensity, which is expensive and could reduce normal efficiency of power extraction.
- The complexity, cost, and uncertainty of the regulatory process are often preventing the full potential of the technology development. Moreover, the new technologies raise concerns such as disturbance or destruction of marine life, disturbance of recreation in nearshore, and threat to navigation from collisions due to the low profile of some wave energy converters above water.

1.2.2 Tidal Power

Tidal energy is another form of ocean renewable energy that can be utilized to power distributed ocean systems. Among various ocean energy technologies under development, tidal turbines are gaining increasing attention because of their efficiency and scalability. Tidal power system is one of the first ocean energy technologies to be commercialized. In 1960s, a large-scale barrage was built in La Rance, France, and small-scale tidal turbines had also been prototyped for research purposes [20, 21]. The kinetic energy from moving water drives a tidal turbine, which is then converted from mechanical energy to electrical energy by a generator. The electricity is then transferred to onshore power station through underwater power cables [22]. In most cases, power electronics are required to condition the power output before interconnection with the grid. The reliability of power electronic devices in renewable energy generation system is discussed in [23]. Permanent magnet synchronous generator (PMSG) is the preferred tidal generator because it shows a very high robustness compared to other types of generator and is driven directly by the turbine hub [24]. The harsh ocean environment in which the tidal turbines are installed and operated leads to severe risks and high costs. They are often exposed to alternating loads in the bottom of the ocean, which is 800 times denser than air. These conditions affect the operations of the tidal energy drivetrain and life cycle of critical components such as the tidal turbine shaft. The power take-off train of a tidal turbine consisting of a shaft is a highly stressed subsystem that works continuously in terms of operational time. At rotating speed of one-third of the speed of the wind turbine under the same power ratings, tidal turbine shaft absorbed larger cyclic stress.

In 2009, an incident occurred in the Bay of Fundy on the Atlantic coast of North America, which resulted in failure of a 1 MW tidal turbine generator shaft after 20 days of installation. The problem was ascribed to high tidal current speed that resulted in a rapid torsional shaft oscillations, thereby destroying the system's rotor blades [25]. Research is ongoing on shaft fatigues in tidal turbine following severe disturbances [26–29].

Reliability is another critical consideration during design and operation of tidal turbine technology. While it is easy for an underwater system to fail under the extreme conditions, it is also difficult to maintain and repair. A reliability model with variable failure rate was developed in [30–33]. The authors used Monte Carlo simulation to generate the failure rate distribution, and 90% confidence interval was determined. However, little work has been published on life cycle extension for tidal turbines.

1.2.2.1   Maximum Power Extraction

In tidal turbine, tidal currents have significant influences on systems. As the complexity of the system installed and operated undersea increases, so does its susceptibility and sensitivity to structural damage and electrical faults. Depending on the scale of the systems and the integration of diverse components, these faults may lead to severe damage to the turbine blades and high costs. Also, tidal turbines which are very similar to wind turbines in terms of their functionality and design, but operated in stochastic underwater ocean environment, are often exposed to hydrodynamic loads. These conditions influence the operations of the tidal energy drivetrain and the life cycle of critical components, such as the turbine shaft [34, 35].

The shaft in a power take-off train of a tidal turbine is a highly stressed subsystem that works continuously in terms of operational time. At a rotating speed of one-third of the speed of the wind turbine under the same power ratings, the tidal turbine shaft absorbs larger cyclic stress [32]. With high-cycle fatigue, the number of cycles and their amplitudes leads to failure, thereby decreasing power production and life span of the system.

Maximum power point tracking (MPPT), a popular adaptive control strategy that extracts the maximum available power in a conversion system, has been used successfully in renewable energy technologies, especially in PV and wind power generation [36, 37]. When applying MPPT in tidal generation system, the reference of the rotational speed control loop is adjusted such that the turbine operates around the maximum power of the tidal speed. However, when the tidal speed exceeds 2.5 m/s and becomes high, the mechanical torque becomes high as well adding to the stress on the turbine.

1.3   Underwater Wireless Power Transfer

Since the seminal work by Tesla in 1893 and 1917 on wireless power transmission using microwaves, there has been significant interest among the scientific and engineering community about the transfer of power without physical connectivity over short as well as long distances. Since Tesla's work, in 1964, Brown demonstrated a mechanism for powering a helicopter using microwave beam. This was followed by the work of Don Otto, who developed in 1971, a small trolley that was powered by induction. In 1973, Los Alamos National Lab demonstrated world's first RFID system.

The field of wireless power transfer (WPT) has been revitalized since the development of resonant induction technique. In 2008, Bombardier and Sony have, respectively, demonstrated wireless transmission product PRIMOVE [38], a wireless charging based on high-power inductive energy transfer. Between 2010 and 2017 multiple coil system, controls and different applications have been demonstrated with moderate efficiency.

In regard to WPT in the ocean environment, it is still very much in the research portion of its development. The challenge lies in the dynamic nature of ocean which keeps the transmitting side of WPT constantly in motion. The transient characteristics and time-varying coupling coefficient of the underwater wireless power transfer (UWPT) are more complex than in air. Furthermore, challenges such as the leakage self-inductance of the transmitting and receiving sides, power capacity of the systems, coil structures, and configurations have to be addressed as well. Previous solution for dynamic WPT in air such as tuning capacitance by using variable capacitor or the well-known impedance control turns out to be very difficult to implement for WPT. Since communication is not viable in an undersea environment, tracking the changes in mutual inductance and coupling coefficient of the system have been particularly challenging.

1.3.1   Wireless Power Transmission Techniques

Wireless power transfer technologies can be extensively grouped into radiative RF-based charging, generally utilized for far-field transmission, and non-radiative coupling-based utilized for near-field transmission. The far-field power transfers are based on using schemes such as the microwave, optical, and acoustic to transfer energy from the transmitter to the receiver. Alternatively, near-field power transfer uses electromagnetic field couplings, such as the inductive, capacitive, and magnetic resonance coupling to transfer energy from the transmitter to the receiver.

1.3.1.1   Microwave

Wireless power transfer using microwave has been investigated since the 1950s [39]. This technology transmits power over long distance, and it is utilized for various applications, e.g., unmanned aerial vehicles (UAVs), electric vehicle, space solar power, and military applications. In the microwave wireless power transfer strategy, the microwave is produced and then passes through a waveguide adapter in order to reduce the outside exposure of the radiation. The microwave transfer by an antenna travels through a medium and then acquired by the receiving dish-like antenna which transfers it back to AC current as shown in Fig. 1.4. In recent years, interest in RF energy-harvesting technology has increased, and microwave wireless power transfer is being actively studied [40]. However, due to the low power transfer efficiency at long distance, the feasibility of using the microwave wireless power transfer method in applications such as AUV charging is under discussion.

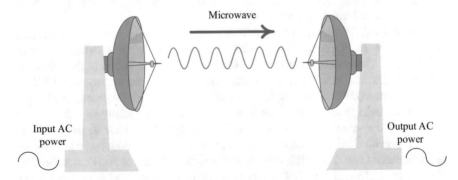

Fig. 1.4 Microwave wireless power transfer

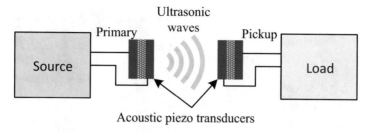

Fig. 1.5 Acoustic wireless power transfer

1.3.1.2 Acoustic

Acoustic WPT has been recently proposed in [41–44]. In the acoustic power transfer, for example, power can be transferred by using piezoelectric transducers to send and receive ultrasonic waves through metal, as shown in Fig. 1.5. This type of acoustic WPT is a feasible approach that can achieve reasonable efficiencies [45]. Moreover, due to the much shorter wavelengths at same frequency as electromagnetic transmitter, acoustic WPT can attain higher beam directive.

However, acoustic WPT method still faces with some challenge. There are significant spatial fluctuations in acoustic intensity due to wave interference, and efficiency quickly drops when the receiver is not centered on the acoustic beams. In particular, the resulting spatial resonance limits the deployment of the receiver and thus limits its application. In addition, the corresponding theoretical investigation is still incomplete, which impedes the development of this method.

1.3.1.3 Optical

Optical power transfer based on laser sources was first introduced for the application of solar power satellite [46]. The concept of optical power transmission using a laser is very similar to microwave power transmission, although, due to its higher efficiency of energy conversion and lesser attenuation, microwave power transfer is more efficient than optical power transfer. However, microwave power transfer requires a very-large-aperture antenna to focus the microwave beam, whereas laser beam in optical power transfer can easily focus in small area, as shown in Fig. 1.6. In addition, compared to other far-field wireless power transfer technologies, optical power transfer shows a salient advantage, such as large coverage, concentrated transmitting energy, long transmission range, and no interference with existing RF applications.

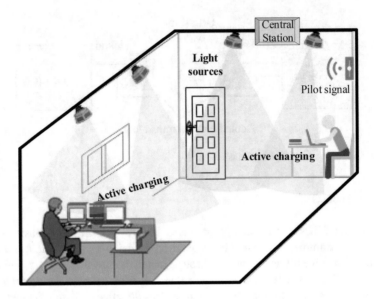

Fig. 1.6 Optical wireless power transfer

1.3.1.4 Capacitive

Capacitive coupling to tube lighting by Tesla in 1891 was the first public testing of WPT to power a load [47, 48]. Capacitive wireless power transfer (CWPT) has been well studied [49] since then. In capacitive wireless power transfer technique, the transmitter and receiver electrodes can form a capacitor, where induced alternating potential of the receiving plate can be generated by the alternating voltage of the transmitting plate through the electrostatic induction. The transmitting power is determined by the very high switching frequency and capacitance between the plate. Most of the present CWPT systems were designed for low power-level applications, e.g., smart lamps, USB devices, and small robots.

1.3.1.5 Inductive

The concept of inductive wireless power transfer is similar to that of a transformer with a weak coupling, i.e., it uses magnetic field induction as a means to transfer energy between two coils. A primary coil energized by a high-frequency AC power source is used as a transmitter, while a secondary coil is used as a receiver.

Typically, an inductive WPT system consists of three parts, transmitter, inductive coupler, and receiver, as seen in Fig. 1.7. The system can be considered to be static when the distance between the transmitting and receiving coil is constant and dynamic when the distance and the load vary. Static inductive WPT for charging devices in air has reached commercialization stage [50], whereas for

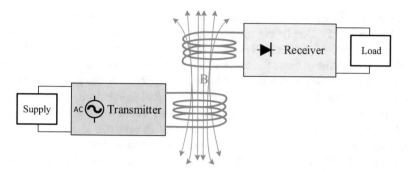

Fig. 1.7 Inductive wireless power transfer

dynamic inductive WPT problems such as the complexity of transient characteristics and time-varying mutual inductance (coupling coefficient) are yet to resolved. Impedance matching circuit has been widely adopted for dynamically changing distances for a WPT in air [51–53].

Recent research has covered different areas such as multiple transmitter and multiple receiver in a dynamic WPT [54], which have the potential to open door to power very low-power devices. However, the authors have reservation on their application potentials for high-power applications due to the limitation of the energy efficiency. For high transfer efficiency, the distance between the transmitting coil and receiving coil is usually less than a wavelength of the frequency, which is in the range of kilohertz. Meanwhile, the conductive material such as the ferrite is also used to direct the magnetic flux and improve the coupling between the transmitting coil and the receiving coil.

1.3.2 Wireless-Powered Ocean System

Distributed ocean technology such as autonomous underwater vehicles (AUVs) operates primarily using the potential energy stored in batteries to perform oceanic research, search and recovery operations, and military purposes. The batteries carry a finite amount of charge and must be replenished in order to continue operations. Conventional charging methods involve removal of the AUV from water, and power is provided through a cable connection. The conductors are primarily copper or aluminum which corrode badly in the presence of salt solutions. Removal of the AUV from the water requires a larger vessel with a crew operating it, increasing the operating costs. In addition, when one buys an AUV, the charger only works with the brand or particular type of vehicle, thereby further increasing the cost of operating multiple AUVs.

To solve these challenges, in the past years, undersea charging docks have been designed, developed, and tested in ocean to charge AUV in-situ. These docking stations include FAU-Morpheus, WHOI-MIT, EDCAOSN, WHOI-REMUS, and ALIVE dock [55]. Due to the unstructured, uncertain, and dynamic nature of the ocean environment, docking stations encounter limitations when deployed in this environment. Some of these limitations include poor maneuverability, susceptibility to biofouling, sensing capability, and high cost. In addition, the design of the dock station allows for dirt, contaminants, and biological agents to be collected in the charging receptacle, which over time, as the dirt buildup, would prevent proper alignment of the coils and impede the function of the system. The most promising approach until now has been the development of an underwater wireless charging station by the US Navy [56]. In July of 2017, researchers from the Naval Surface Warfare Center transferred 2 kW of electricity from a submerge wireless charging station into an AUV. However, the method used requires the AUV to land on a dock through a ship and controlled to plug into it, which limit its capabilities on remote autonomous operations.

The wireless transfer of power to the batteries while the AUV is underwater provides a solution to these operational complications. Furthermore, automating the recharging process using a wave energy converter (WEC) increases the independence of the AUV and will prolong operational periods while decreasing maintenance expenditures.

1.3.2.1 Challenges

The WPT system will be operating submerged in ocean water and subject to the force of waves, temperature fluctuation, and corrosive salts. Hence, the design must be watertight to prevent intrusion onto the conductors and rugged enough to withstand constant motion of the waves. Summarized below are other challenges facing WPT system in underwater environment.

- Seawater itself has inherent properties which make transmission of electromagnetic waves difficult due to its conductance. As seen later in Chap. 3, the salt increases conductivity and provides paths for eddy currents to form within the transfer path. Eddy currents create opposing magnetic fields to the origin of their flux and diminish the field strength of the WPT system coils.
- Based on the analysis provided in this book, the radiation resistance of sea water is increased rapidly above frequencies of 200 kHz and would dominate the DC and AC resistances at higher frequencies. Radiation resistance often limits operating frequency of the primary voltage and needs to remain below the threshold to maintain efficiency.
- Another constraint that must be considered is the physical dimensions of the transmitting and receiving coils in relation to the AUV and devices being charge. Too large of coils will be cumbersome for integrating into existing systems, and too small of coils will limit the throughput of power. The ratio of optimum coil size must be derived through simulation and experimental measurement.

1.4 Summary

This chapter is about a brief review of existing wave and tidal technologies and the underwater wireless charging techniques for ocean systems. The challenges such as irregular wave motion, power intermittent, and powering distributed ocean systems in the harsh ocean environments are reviewed. A Smart-WEC, an isolated remote ocean energy source which could provide a sustainable power solution for developing ocean spaces, is presented. This technology will be particularly advantageous for distributed ocean system applications like marine monitoring sensor suites, whose data collection can be hindered by limited power, and autonomous underwater vehicles (AUVs), whose mission range is often curtailed by battery power limitations.

References

1. J. Raja, L.C. Videira, B. Pierre, Battery lifetime estimation and optimization for underwater sensor networks. IEEE Sensor Netw. Oper. **2006**, 397–420 (2004)
2. M. Annette, J.G. Vining, Ocean wave energy conversion-a survey, in *Industry Applications Conference, 41st IAS Annual Meeting*, vol. 3 (2006), pp. 1410–1417
3. J. Lehmkoster, T. Schroder, D. Ladischensky, Marine minerals and energy. Tech. Rep. 2010 [Online]. http://worldoceanreview.com/wp-content/downloads/wor1/WOR1chapter7.pdf
4. E. Siirila, Un atlas: 44 percent of us live in coastal areas [Online]. https://coastalchallenges.com/2010/01/31/un-atlas-60-of-us-live-in-the-coastal-areas/
5. C. João, S. Rebecca, S. Philip, T.R. Eatock, Estimating the loads and energy yield of arrays of wave energy converters under realistic seas. IET Renew. Power Gen. **4**(6), 488 (2010)
6. Z. Peng, W. Yang, X. Weidong, L. Wenyuan, Reliability evaluation of grid-connected photovoltaic power systems. IEEE Trans. Sustain. Energy **3**(3), 379–389 (2012)
7. W. Yang, Z. Peng, L. Wenyuan, K. Nadim, Comparative analysis of the reliability of grid-connected photovoltaic power systems, in *IEEE Power and Energy Society General Meeting*, San Diego (2012), pp. 1–8
8. J. Zhang, X. Xiao, P. Zhang, J. Lu, T. Orekan, Subsynchronous control interaction analysis and a trigger-based active damping control for dfig-based wind turbines. Electr. Power Compon. Syst. **44**(7), 713–725 (2016)
9. D. Benjamin, A.R. Plummer, S.M. Necip, A review of wave energy converter technology. Proc. Inst. Mech. Eng. A J. Power Energy **223**(8), 887–902 (2009)
10. H. Titah-Benbouzid, M. Benbouzid, Development and demonstration of the WEC-sim wave energy converter simulation tool, in *Proceedings of the 2nd Marine Energy Technology Symposium*, Shanghai, China (2014)
11. H. Titah-Benbouzid, M. Benbouzid, An up-to-date technologies review and evaluation of wave energy converters. Int. Rev. Electr. Eeng. IREE **10**(1), 52–61 (2015)
12. P. Holmberg, M. Andersson, B. Bolund, Kerstinand, T. Schroder, D. Ladischensky, Wave power surveillance study of the development. Tech. Rep. 2011 [Online]. https://energiforskmedia.blob.core.windows.net/media/19924/wave-power-surveillance-study-of-the-development-elforskrapporter-2011-02.pdf. Accessed 10 June 2018
13. NREL.GOV, New wave energy converter design inspired by wind energy. Tech. Rep., 2018 [Online]. https://www.nrel.gov/news/program/2018/new-wave-energy-converter-design-inspired-by-wind-energy.html. Accessed 7 Sept. 2018
14. P.B. Garcia-Rosa, J.P.V.S. Cunha, F. Lizarralde, S.F. Estefen, I.R. Machado, E.H. Watanabe, Wave-to-wire model and energy storage analysis of an ocean wave energy hyperbaric converter. IEEE J. Ocean. Eng. **39**(2), 1817–1828 (2014)

15. P.C.J. Clifton, A. McMahon, H.P. Kelly, Design and commissioning of a 30kw direct drive wave generator, in *IET Conference on Power Electronics, Machines and Drives (PEMD)*, Brighton, UK (2010)
16. L. Cappelli, F. Marignetti, G. Mattiazzo, E. Giorcelli, G. Bracco, S. Carbone, C. Attaianese, Linear tubular permanent-magnet generators for the inertial sea wave energy converter. IEEE Trans. Ind. Appl. **50**(3), 1817–1828 (2014)
17. D.E.A.M. Andrade, A. de la Villa Jaén, A.G. Santana, Improvements in the reactive control and latching control strategies under maximum excursion constraints using short-time forecast. IEEE Trans. Sustain. Energy **7**(1), 427–435 (2016)
18. Z. Feng, E.C. Kerrigan, Latchingdeclutching control of wave energy converters using derivative-free optimization. IEEE Trans. Sustain. Energy **6**(3), 773–780 (2015)
19. F. Fusco, J.V. Ringwood, A simple and effective real-time controller for wave energy converters. IEEE Trans. Sustain. Energy **4**, (1), 21–30 (2015)
20. M. Rowell, Experimental evaluation of mixer ejector hydrokinetc turbine (MEHT) at two tidal energy test sites and in a tow tank. Tech. Rep. 2013 [Online]. http://www.mrec.umassd.edu/media/supportingfiles/mrec/agendasandpresentations/4thconference/matthew/rowell.pdf
21. M. Shahsavarifard, Effect of shroud on performance of horizontal axis hydrokinetic turbine. Tech. Rep. 2013 [Online]. http://www.mrec.umassd.edu/4thconference
22. Y. Zhao, X. Su, Tidal energy: technologies and recent developments, in *IEEE International Energy Conference and Exhibition (EnergyCon)* (2013), pp. 618–623
23. W. Huai, M. Liserre, F. Blaabjerg, R.D. Place, J. Jacobsen, T. Kvisgaard, J. Landkildehus, Transitioning to physics-of-failure as a reliability driver in power electronics. IEEE J. Emerg. Sel. Topics Power Electron. **2**(1), 97–114 (2014)
24. S. Benelghali, M.E.H. Benbouzid, J. Charpentier, T. Ahmed-Ali, I. Munteanu, Experimental validation of a marine current turbine simulator: application to a permanent magnet synchronous generator-based system second-order sliding mode control. IEEE Trans. Ind. Electron. **58**(1), 118–126 (2011)
25. K. Sean, Failed tidal turbine explained at symposium [Online]. http://www.cbc.ca/news/canada/nova-scotia/failed-tidal-turbine-explained-at-symposium-1.1075510
26. D.N. Walker, S.L. Adams, R.J. Placek, Torsional vibration and fatigue of turbine-generator shafts. IEEE Trans. Power Apparatus Syst. **58**(11), 4373–4380 (1981)
27. M. Jackson, S. Umans, R. Dunlop, S. Horowitz, A. Parikh, Turbine-generator shaft torques and fatigue: Part I - simulation methods and fatigue analysis. IEEE Trans. Power Apparatus Syst. **98**(6), 2299–2307 (1979)
28. T. Hammons, Accumulative fatigue life expenditure of turbine/generator shafts following worst-case system disturbances. IEEE Trans. Power Apparatus Syst. **101**(7): 2364–2374 (1982)
29. J. Song-Manguelle, S. Schroder, T. Geyer, G. Ekemb, J. Nyobe-Yome, Prediction of mechanical shaft failures due to pulsating torques of variable-frequency drives. IEEE Trans. Ind. Appl. **46**(5), 979–1988 (2010)
30. C. Iliev, V. Val, Tidal current turbine reliability: power take-off train models and evaluation, in *3rd International Conference on Ocean Energy* (2010)
31. D.A. Douglas, T. Brekken, Monte carlo analysis of the impacts of high renewable power penetration, in *IEEE Energy Conversion Congress and Exposition (ECCE)* (2011), pp. 3059–3066
32. J. Song-Manguelle, S. Schröder, T. Geyer, G. Ekemb, J.-M. Nyobe-Yome, Prediction of mechanical shaft failures due to pulsating torques of variable-frequency drives. IEEE Trans. Ind. Appl. **46**(5), 1979–1988 (2010)
33. A. Secil, Fatigue life calculation by rainflow cycle counting method. Master's thesis, Middle East Technical University, Ankara, Turkey, 2004
34. D.N. Walker, S.L. Adams, R.J. Placek, Torsional vibration and fatigue of turbine-generator shafts. IEEE Trans. Power Apparatus Syst. **PAS-100**(11), 4373–4380 (1981)
35. C. Iliev, D. Val, Tidal current turbine reliability: power take-off train models and evaluation, in *Proceedings of 3rd International Conference on Ocean Energy*, Bilbao (2010)

36. S. Adhikari, L. Fangxing, Coordinated v-f and p-q control of solar photovoltaic generators with mppt and battery storage in microgrids. IEEE Trans. Smart Grid **5**(3), 1270–1281 (2014)
37. M. Pucci, M. Cirrincione, Neural MPPT control of wind generators with induction machines without speed sensors. IEEE Trans. Ind. Electron. **58**(1), 37–47 (2010)
38. A. Brecher, D. Arthur, Review and evaluation of wireless power transfer (WPT) for electric transit applications. Tech. Rep. 2014 [Online]. https://www.transit.dot.gov/sites/fta.dot.gov/files/FTA_Report_No._0060.pdf
39. W.C. Brown, The history of power transmission by radio waves. IEEE Trans. Microw. Theory Tech. **32**(9), 1230–1242 (1984)
40. S. Sasaki, K. Tanaka, K. Maki, Microwave power transmission technologies for solar power satellites. Proc. IEEE **101**(6), 1438–1447 (2013)
41. T. Ishiyama, Y. Kanai, J. Ohwaki, M. Mino, Impact of a wireless power transmission system using an ultrasonic air transducer for low-power mobile applications, in *IEEE Symposium on Ultrasonics, 2003*, vol. 2 (2003), 1368–1371
42. M.G.L. Roes, M.A.M. Hendrix, J.L. Duarte, Contactless energy transfer through air by means of ultrasound, in *IECON 2011 - 37th Annual Conference of the IEEE Industrial Electronics Society* (2011), pp. 1238–1243
43. M.G.L. Roes, J.L. Duarte, M.A.M. Hendrix, E.A. Lomonova, Acoustic energy transfer: a review. IEEE Trans. Ind. Electron. **60**(1), 242–248 (2013)
44. T.C. Chang, M.J. Weber, M.L. Wang, J. Charthad, B.P.T. Khuri-Yakub, A. Arbabian, Design of tunable ultrasonic receivers for efficient powering of implantable medical devices with reconfigurable power loads. IEEE Trans. Ultrason. Ferroelectr. Freq. Control **63**(10), 1554–1562 (2016)
45. V.F. Tseng, S.S. Bedair, N. Lazarus, Phased array focusing for acoustic wireless power transfer. IEEE Trans. Ultrason. Ferroelectr. Freq. Control **65**(1), 39–49 (2018)
46. P.E. Glaser, Power from the sun: its future. Science **162**(3856), 857–861 (1968)
47. N. Tesla (ed.), *Experiments With Alternate Currents of Very High Frequency and Their Application to Methods of Artificial Illumination* (Wilder, Radford, 2008)
48. T.C. Martin, N. Tesla (eds.), *The Inventions Researches and Writings of Nikola Tesla With Special Reference to His Work in Polyphase Currents and High Potential Lighting* (Prabhat, New York, 1894)
49. J. Dai, D.C. Ludois, A survey of wireless power transfer and a critical comparison of inductive and capacitive coupling for small gap applications. IEEE Trans. Power Electron. **30**(11), 6017–6029 (2015)
50. M.H.M. Salleh, N. Seman, D.N.A. Zaidel, Design of a compact planar witricity device with good efficiency for wireless applications, in *2014 Asia-Pacific Microwave Conference*, Sendai (2014), pp. 139–137
51. H.W. Benjamin, P.S. Alanson, R.S. Joshua, Adaptive impedance matching for magnetically coupled resonators, in *PIERS Proceedings*, Moscow (2012), pp. 694–702
52. L. Yongseok, T. Hoyoung, L. Seungok, P. Jongsun, An adaptive impedance-matching network based on a novel capacitor matrix forwireless power transfer. IEEE Trans. Power Electron. **29**(8), 4403–4414 (2014)
53. E.K. Kim, C.B. Teck, I. Takehiro, H. Yoichi, Impedance matching and power division using impedance inverter for wireless power transfer via magnetic resonant coupling. IEEE Trans. Ind. Appl. **50**(3), 2061–2071 (2014)
54. X. Yu, T. Skauli, B. Skauli, S. Sandhu, P. Catrysse, S. Fan, Wireless power transfer in the presence of metallic plates: experimental results. AIP Adv. **3**(6), 062102 (2013)
55. H. Singh, S. Lerner, K. von der Heyt, B.A. Moran, An intelligent dock for an autonomous ocean sampling network, in *IEEE Oceanic Engineering Society. OCEANS'98. Conference Proceedings (Cat. No.98CH36259)*, vol. 3 (1998), pp. 1459–1462
56. M. Luciano, Navy's underwater wireless charging station can improve remote UUV mission performance (2018) [Online]. https://www.ecnmag.com/blog/2018/05/navys-underwater-wireless-charging-station-can-improve-remote-uuv-mission-performance. Accessed 11 Aug. 2018

Chapter 2
Design and Modeling of a Smart Wave Energy Converter

2.1 Introduction

Currently, several types of WEC concepts are under investigation [1]. The WEC has yet to reach the stage of widespread deployment of commercial devices that other renewable energy, such as wind and solar power have attained. For the wave energy technologies to withstand the ocean's harsh environment, hydrodynamic assessments of the device performance are critically important. For the sake of simplicity, some ongoing research focusing on the hydrodynamic models of WECs assumes linear coefficients, and their performance evaluations are often based on the sinusoidal assumption for incident waves [2]. In practice, WEC is not always operating under the single sine wave patterns. WEC can only be correctly described with its complex nonlinear equations, which in most cases are difficult to parameterize accurately. In [3], the optimal discrete power take-off force is investigated for a simplified WEC; however, the method failed for nonlinear irregular waves input.

Several challenges need to be tackled to successfully create a WEC that exploit the energy from waves and operate at high-energy conversion efficiency, especially under in extreme weather conditions where loading on the device can be 100 times higher than average loading. Of particular difficulty in WEC technology, such as the direct drive point absorber, is the coupling of the slow motion (0.1 Hz) to drive a linear generator, with an acceptable power output quality.

In this chapter, a smart wave energy converter (Smart-WEC), a new type of power harvester that can extract energy from the motion of the ocean waves, is presented. It has the potential to resolve the long-standing challenge of providing sustainable power to distributed ocean applications while addressing the limitations of conventional ocean technology to allow a more robust and reliable system. Compared to existing WEC technology [4], the Smart-WEC has the capability of making wireless power connections with underwater ocean technologies through our novel UWPT technology [5]. The proposed technology will make a profound

© The Author(s), under exclusive licence to Springer Nature Switzerland AG 2019
T. Orekan, P. Zhang, *Underwater Wireless Power Transfer*, SpringerBriefs in Energy,
https://doi.org/10.1007/978-3-030-02562-5_2

impact on the development of storm monitoring systems such as marine automated network buoys, Tsunami buoys, and autonomous underwater charging stations. An integrated nonlinear wave-to-wire model of the device in regular and irregular wave conditions is developed. The modeling results help us better understand the sensitivity of WEC system parameters, identify several unique design constraints that have not previously been explored, and develop a reliable control to increase power production.

2.1.1 Conceptual Design of the Smart-WEC

As can be seen in the conceptual 3D structure provided in Fig. 2.1, the proposed Smart-WEC is a single body point absorber wave energy converter [6]. It consists of a buoy oscillating in heave at the ocean's surface, connected to a relatively stationary cylindrical spar fixed to the sea bed (near shore), and a PTO mechanism. Currently, the most common PTO is the hydraulic unit PTO, which converts the low-frequency and high-force motion of the WEC into a high-speed motion to produce power. However, this type of PTO requires more moving parts and could require more regular maintenance, resulting in very high maintenance cost. In the presented Smart-WEC, energy is extracted from the motion of the buoy through a direct drive linear generator (DDLG). The DDLG is a lightweight air-cored linear generator. Unlike iron-cored LGs, air-cored LGs generally have a much lower inductance [7]. Additionally, their ability to eliminate attraction and cogging forces makes air-cored

Fig. 2.1 Smart-WEC
conceptual structure

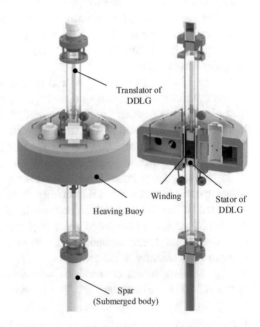

Translator of
DDLG

Winning   Stator of
DDLG

Heaving Buoy

Spar
(Submerged body)

models a better option. That being said, the internal resistance of the presented air-cored DDLG is more significant due to the large number of coil turns needed to increase the voltage induced. Furthermore, the cylindrical buoy is restricted to 1-DoF (heave motion), and the stroke length of the translator is designed for real wave height, allowing for increased power output by the DDLG.

Compared to the WEC design currently being developed, the presented Smart-WEC [5, 8, 9] enables wireless powering of distributed ocean systems, through a novel underwater wireless power technology.

2.2 Mathematical Modeling of WEC

Mathematical models for designing and controlling WECs are indispensable. While considering the nonlinear terms improve the accuracy of the model, it also requires additional computational resources. Typically, the mathematical approach follows the well-known Cummin's integro-differential equation [10], to analyze the interactions of the system in marine environment. Most studies use software like WAMIT for hydrodynamic parameter identification of the system.

2.2.1 Time and Frequency Domain Method

In choosing a modeling method for WEC, the parameters should always either represent the radiation force convolution term or the system's dynamic in terms of force to motion.

2.2.1.1 Time Domain Formulation

Using Newton's second law, the equation of motion for a 1-DoF WEC can be described in time domain as

$$m\ddot{z}(t) = f_e(t) + f_r(t) + f_h(t) + u(t) \tag{2.1}$$

where m is the mass of the buoy, $z(t)$ the device heave excursion, $f_e(t)$ the wave excitation force, $f_r(t)$ the wave radiation force, and $f_h(t)$ the hydrostatic restoring force which can be written as $-s_h x(t)$. The $u(t)$ is the control input supplied by a power take-off (PTO) system. The $f_r(t)$ is modeled from linear potential theory in [10]. Boundary elements method is usually adopted to generate the impulse response data. The completed linearized equation of motion of the WEC is given by

$$(m + u_\infty)\ddot{z}(t) = k_t(t) * \dot{z}(t) + s_h x(t) = f_e(t) + u(t) \tag{2.2}$$

where u_∞ is the added mass at infinite frequency and $k(t)$ is the convolution kernel.

2.2.1.2 Frequency Domain Formulation

While considering velocity as the output, applying the Fourier transform to Eq. (2.2) is given by

$$\hat{\dot{x}}(j\omega) = \hat{F}_e(j\omega)H(j\omega) \tag{2.3}$$

Here $H(j\omega)$, is a function of a particular set of frequency-dependent parameters, as represented

$$H(j\omega) = \frac{1}{b_u + B(\omega) + j\omega[A(\omega) + m + m_u] + \frac{s_h + s_u}{j\omega}} \tag{2.4}$$

where $A(\omega)$ and $B(\omega)$ represent the radiation added mass and damping of the WEC [11]. These two parameters should be efficiently obtained. An example is given in [12], to improve the data obtain when using hydrodynamic solvers, such as WAMIT, to compute the parameters $A(\omega)$ and $B(\omega)$.

2.2.2 Nonlinear Effects in WEC

Much of the existing literature deals with control at the mechanical power take-off levels, referencing the generator control as a system input. In the hydrodynamic modeling, each of the force exerted on a WEC can be a source of nonlinearities [13], and it is essential to see how they impact the WEC. All the forces acting on a WEC can be described as

$$M\ddot{Z}(t) = f(F_g(t), F_{FK}(t), F_{diff}(t), F_{add}(t), F_{moor}(t), F_{PTO}(t), F_{vis}(t), F_r(t) \tag{2.5}$$

where F_{FK} is the Froude-Krylov force, F_{diff} is the diffraction force, F_{add} and F_g are the additional force and gravity force, F_{moor} is the mooring force, F_{PTO} is the acting force on WEC due to the PTO system, and F_{vis} is the viscous force.

Except for the gravity and added force, other forces are analyzed as follows:

- *Froude-Krylov force F_{FK}*: represents the integral of the static and dynamic pressure on the WEC device surface. The dynamic parts represent the incoming incident waves, while the static part represents the relation between the system's buoyancy and gravity forces. The influence of the nonlinear F_{FK} plays an important role, as the power absorption is often overestimated when F_{FK} is not considered [13].
- *Diffraction force F_{diff}*: is the load observed during diffracted wave and causes disturbances which is introduced to the WEC by the presences of the floating bodies. Considering this term in the nonlinear equation generally improves the accuracy of the model.

- *Mooring force F_{moor}*: often applied to keep the WEC device in position, when subject to drift forces due to tides, waves, and wind. The design of moorings can be very challenging because of the nonlinear behavior during survival waves. Without consideration of the mooring nonlinear phenomena in WEC design, the system may fail or result in significant reduction in power efficiency [13].
- *PTO force F_{PTO}*: the PTO is part of the WEC device that utilizes its relative motion to produce power. In a direct drive linear generator PTO, as the translator of the generator moves against the armature and induced a magnetic induction, some forces due to this interaction can lead to nonlinear behavior.
- *Viscous force F_{vis}*: the existence of the nonlinear viscous force, which can be identified when using wave tank experiments, is crucial to the dynamic response of the device. In general, the importance of viscous force effects depends on the type, shape, and dimension of the WEC.

2.3   Modeling of the Smart-WEC

The number of freedom of the motion of a WEC structure is a key issue that determines its optimal operation. The Smart-WEC has 6-DoF, the heave, yaw, sway, pitch, roll, and surge as shown in Fig. 2.2. Other degrees of freedom could be considered in the case of extreme wave conditions. Little literature has been studied utilizing 6-DoF WEC motion to reach the maximum utilization of the wave energy available and improve the energy harvest capabilities of the system. In [14], the dynamic and kinematic analyses of multiple DoF mechanism for WEC have been studied.

The Smart-WEC device is modeled by calculating the solution for its dynamic equations under the assumption that the device can only move in the heave motion,

Fig. 2.2 Smart-WEC six
degree of freedom

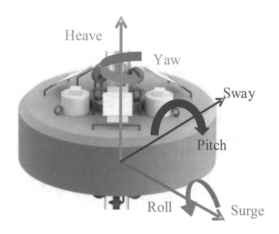

while other DOFs are restricted. The dynamic equation may be written according to Newton's second law which describes bodies in motion with a single degree of freedom as

$$M\ddot{z}(t) = f_e(t) + f_r(t) + f_v(t) + f_b(t) + f_{gen}(t) + f_m(t) \tag{2.6}$$

where $\ddot{z}$ is the (heave excursion) acceleration, M is the mass of the buoy, $f_e(t)$ is the hydrodynamic excitation force, $f_r(t)$ is the wave radiation force due to radiated waves when the body moves, $f_v(t)$ is the viscous damping force, $f_b(t)$ is the buoyancy force, $f_{gen}(t)$ is the force produce by the DDLG, and $f_m(t)$ is the force due to mooring connection. Although, as discussed in Sect. 2.2.2, the effect of viscous force should be considered in WEC design. However, in [13], the vicious effects is said to have low influence on the behavior of a small heaving point absorber and its power production. Hence, the influence of viscosity and mooring will not be included in the Smart-WEC equation of motion, only the forces due to the heave motion of the device will be considered. Equation (2.6) can be rewritten as a partially nonlinear equation, given by

$$M\ddot{z} = f_e(t) + f_r(t) + f_b(t) + f_{gen}(t) \tag{2.7}$$

The wave radiation force $f_r(t)$ can be expressed by the linear convolution of the radiation impulse response function $K(t)$ and the added mass term $m_{a,\infty}\ddot{z}(t)$, as shown in Eq. (2.8). The added mass $m_{a,\infty}$ at infinite wave frequency represents the inertia of the surrounding fluid, which is added to the device thereby increasing its inertia.

$$f_r(t) = -\int_0^t K_r(t - \tau)\dot{z}(t)(\tau)d\tau - m_{a,\infty}\ddot{z}(t) \tag{2.8}$$

$$f_b(t) = -\rho g S \int_0^t z(\tau)d\tau = -K_b z(t) \tag{2.9}$$

Equation (2.9) is the buoyancy force $f_b(t)$ that models the hydrostatic force related to the WEC'S displacement through a hydrostatic stiffness constant K_b. The force $f_b(t)$ depends on the water density ρ, acceleration due to gravity g, and cross-sectional surface area S of the device. Thus, the hydrodynamics of the Smart-WEC response can be expressed as an integro-differential equation, based on Cummin's decomposition [15], which results in

$$(M + m_{a,\infty})\ddot{z}(t) = -\int_0^t K_r(t - \tau)\dot{z}(t)(\tau)d\tau - K_b z(t)$$
$$+ f_{gen}(t) + f_e(t) \tag{2.10}$$

2.4 Electrical Model

The electrical system of the Smart-WEC is formed by three main subsystems, which consist of the DDLG generator, the power electronic converters, and the battery energy storage (Fig. 2.3). The PTO mechanism, DDLG, is similar in principle to the direct drive permanent magnet generator used in wind energy technology where the moving part of the generator is directly coupled with the energy-absorbing part. It consists of a set of neodymium-iron-boron permanent magnets with alternating polarity. The DDLG is designed to take advantage of the slow movement of the ocean waves that is transferred to it through the heaving buoy. It operates under variable speeds depending on the conditions of the sea waves. Since the DDLG is air-cored, the inductance has very little effect (considered negligible) on the low-frequency current, thereby, allowing current to be controlled independently. This characteristic becomes an important feature in the presented MPEC discussed in Chap. 5. System parameters can be found in Table 2.1.

In the three-phase DDLG, the coils are displaced horizontally by a third of the magnetic wave length due to the reciprocation motion of the translator. The dq model of the three-phase-induced EMFs can then be written as

$$e_d = -R_s i_d + \frac{d\lambda_d}{d(t)} - \omega\lambda_q \tag{2.11}$$

$$e_q = -R_s i_q + \frac{d\lambda_q}{d(t)} - \omega\lambda_d \tag{2.12}$$

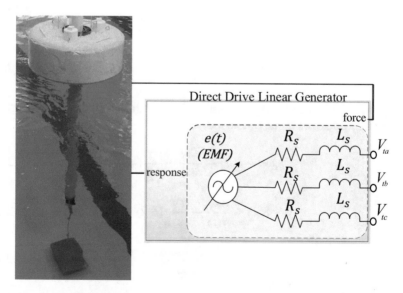

Fig. 2.3 Smart-WEC prototype: electrical system model of the DDLG

Table 2.1 Simulation
parameters

Parameter	Value
Wave height	2.5 m
Pole pitch	6.5 mm
Wave time period	8 s
Phase number	3
Buoy mass	50 Kg
Filter inductance	1.8 mH
Rated voltage	300 V
Switching frequency	25 KHz
Rated current	5 A
Filter capacitance	5 μF
Rated power	2 kW
Phase winding resistance	7.69 Ω
Air-gap length	1 mm
Scaling factor	0.065

where ω is the translator velocity. e_d, e_q, i_d, i_q, λ_d, and λ_q are the EMF voltage, stator currents in the direct and quadrature axes, and flux linkage components in the dq reference frame, respectively. R_s and L_s are the resistance of the stator and the synchronous inductance, respectively. Where λ_{pm} is the flux linkage of the stator winding due to permanent magnet, λ_d and λ_q can be written as

$$\lambda_d = -L_s i_d + \lambda_{pm} \tag{2.13}$$

$$\lambda_q = -L_s i_q \tag{2.14}$$

and

$$\omega = \frac{\pi v}{\tau_p} \tag{2.15}$$

Considering that ω can be expressed using Eq. (2.15), the dynamic model of the DDLG in dq reference frame can be rewritten as

$$e_d = -R_s i_d + \frac{d\lambda_d}{d(t)} + \frac{\pi v}{\tau_p} L_s i_q \tag{2.16}$$

$$e_q = -R_s i_q + \frac{d\lambda_q}{d(t)} + \frac{\pi v}{\tau_p} \left(L_s i_d - \lambda_{pm} \right) \tag{2.17}$$

$$f_{gen} = \frac{3\pi \lambda_{pm}}{2\tau_p} i_q \tag{2.18}$$

where v is the linear speed of the translator and τ_p is the pole pitch and f_{gen} is the generator force, respectively. Note that, DDLG is designed without the need for a linear position, i.e., the area of the stator covered by the translator is not dependent on the translator's position and is assumed to be constant.

2.4.1 Incoming Waves

The energy from the wind is transferred to the ocean in the form of waves and moved across its surface. These incoming waves can be represented as regular sinusoidal waves and nonlinear irregular waves.

2.4.1.1 Regular Waves

Regular waves, as described in Eq. (2.19), consist of a sinusoidal single frequency wave. These type of waves do not exist in reality but can be approximated and define by an amplitude, a frequency, and a phase.

$$\eta = A\cos(\omega t + \phi) \tag{2.19}$$

2.4.1.2 Irregular Waves

A more realistic waves are the irregular waves described in Eq. (2.20). It combines regular waves with different amplitudes, frequencies, and phases. The most established way of describing real sea state is using Pierson-Moskowitz spectrum or the JONSWAP, to analyze records of waves taken at a site [16].

$$\eta = \sum_{i=1}^{N} H_i \cos(\omega_i t + \phi_i) \tag{2.20}$$

2.5 Test Cases

In order to better illustrate the applicability of the implemented hydrodynamic models, this section presents the time domain simulations of the Smart-WEC. Simulations are carried out under regular waves and under more realistic irregular wave conditions. Also, the hydrodynamic parameters of the device and forces such as K_r and f_e are numerically calculated using the WAMIT® program [17], a BIEM method. In addition, a WEC-Sim toolbox, developed by the National Renewable Energy Laboratory and the Sandia National Laboratory, is integrated with the DDLG model implemented in MATLAB® Simulink, to solve the Smart-WEC system dynamics. The DDLG is connected to the rectifier in parallel with an LC filter using the design parameters shown in Table 2.1.

2.5.1 Regular Waves

The characteristics of the Smart-WEC system in regular wave conditions are studied. Simulations are run with a regular wave height of $H_s = 2.5$ m and a period of 8 s. Figure 2.4a, b show the regular wave profile and the plot of the Smart-WEC response. The position, velocity, and acceleration of the device can be seen in the figure. The excitation, frictional, and PTO forces, shown in Fig. 2.4c, d, have been obtained by approximation using the transfer functions of the Smart-WEC frequency response. Shown in Fig. 2.5 are the results of power absorbed, electrical power, and terminal voltage of the DDLG. The average power absorbed and electrical power at the DDLG terminal, as seen in Fig. 2.5a, are 266 W and 170 W, respectively.

2.5.2 Irregular Waves

It is essential to analyze the behavior and performance of the Smart-WEC device under more realistic irregular wave conditions. To reproduce the irregular waves of the sea state shown in Fig. 2.6a, the widely adopted JONSWAP spectrum is used [16] with a significant wave height of 2.5 m and peak period of 8 s. In this test case, compared to the Smart-WEC response when subject to regular waves, the following conclusions can be drawn:

- The results in Fig. 2.6b, c, and d show that there is a significant difference in the device excitation force, buoy responses, and nonlinear PTO force when subjected to irregular waves. This also indicates that when only regular waves and linear instantaneous forces are considered while modeling a WEC, the average power absorbed by the system could be overestimated.
- As seen in Fig. 2.7a, when the device is subjected to realistic irregular wave conditions, it causes large fluctuations in power absorption, indicating that there is a major difference in the power computed using regular waves. The most noticeable difference is the peak-to-average power ratio which exceeds other renewable energy technologies. It is evident from these observations that without proper control, these slow-power fluctuations may cause significant thermal cycling of the system power electronic converters, which would have to be considerably overrated for such system.
- Figure 2.7b shows the three-phase terminal voltage of the DDLG. From the results we can see the effect of the reciprocating wave motion on the voltage.

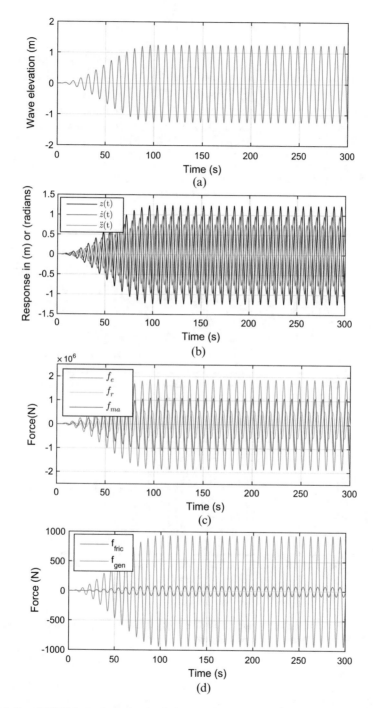

Fig. 2.4 Smart-WEC hydrodynamic simulation results, (**a**) regular wave elevation, (**b**) buoy response (position, velocity, and acceleration), (**c**) buoy excitation and radiation, forces (**d**) frictional and generator forces

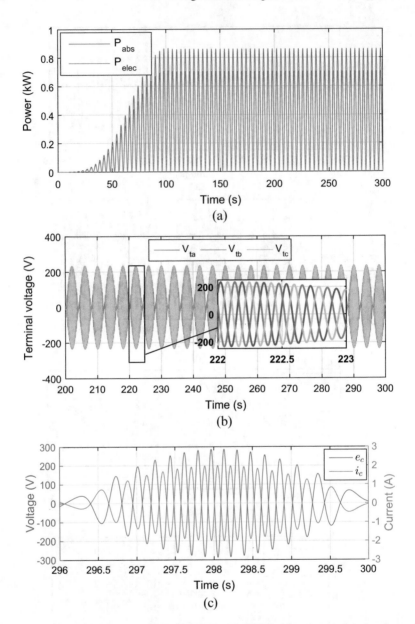

Fig. 2.5 (**a**) Power absorbed by the Smart-WEC, (**b**) DDLG three-phase terminal voltage, (**c**) induced EMF voltage out of phase with current

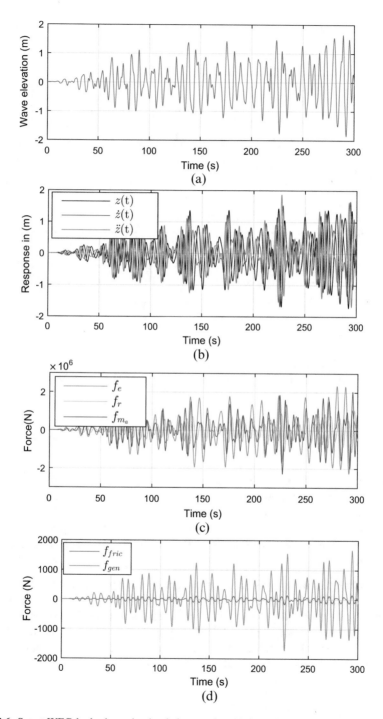

Fig. 2.6 Smart-WEC hydrodynamic simulation results, (**a**) irregular wave elevation, (**b**) buoy response (position, velocity, and acceleration), (**c**) buoy excitation and radiation forces, (**d**) frictional and generator forces

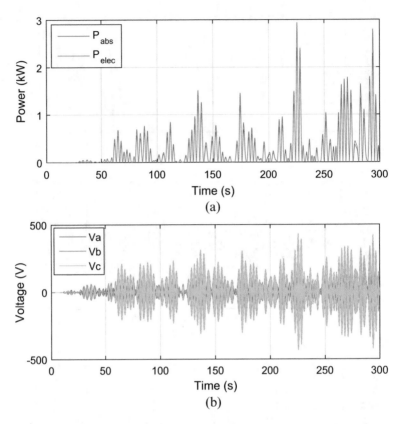

Fig. 2.7 (**a**) Power absorbed and electrical power, (**b**) DDLG three-phase terminal voltage

Within each cycle there are zero dips, and a power electronic rectifier is required to enable constant power flow to the load. The average power absorbed is $P_{abs} = 341$ W, almost 75 W more when the system is subject to regular waves. The power at the generator terminal which is transferred to the power electronics, without control as seen in Fig. 2.7a, is $P_{elec} = 211$ W; this results in an efficiency of 62%.

In the irregular wave conditions, the peak-to-average power ratio and the average power absorbed by the system is higher than that of the regular waves. The rating of the linear generator and the power electronics equipment should also be taken into consideration when designing the MPEC controller. Also, for an iron-cored Smart-WEC DDLG, the effect of the higher inductance in iron-cored LG will have to be considered when MPEC is adapted.

2.6 Summary

A detailed partially nonlinear hydrodynamics model of Smart-WEC device, integrated with a DDLG, has been evaluated under different wave conditions. The Smart-WEC system performances obtained from the hydrodynamic model, when tested in regular and irregular waves, show that there is a large fluctuation in the power absorbed. The excitation and the radiation force calculations are relevant to capturing the nonlinear phenomena of the device.

References

1. B. Czech, P. Bauer, Wave energy converter concepts: design challenges and classification. IEEE Ind. Electron. Mag. **6**(2), 4–16 (2012)
2. L. Cappelli, F. Marignetti, G. Mattiazzo, E. Giorcelli, G. Bracco, S. Carbone, C. Attaianese, Linear tubular permanent-magnet generators for the inertial sea wave energy converter. IEEE Trans. Ind. Appl. **50**(3), 1817–1828 (2014)
3. H.H. Anders, F.A. Magnus, B. Michael, Model predictive control of a wave energy converter with discrete fluid power power take-off system. Energies **11**(3), 635 (2018)
4. H. Titah-Benbouzid, M. Benbouzid, An up-to-date technologies review and evaluation of wave energy converters. Int. Rev. Electr. Eng. IREE **10**(1), 52–61 (2015)
5. T. Orekan, P. Zhang, C. Shih, Analysis, design and maximum power efficiency tracking for undersea wireless power transfer. IEEE J. Emerg. Sel. Topics Power Electron. **6**(2), 843–854 (2017)
6. T. Orekan, P. Zhang, Integrated dynamic modeling of a point absorber wave energy converter in regular and irregular waves, in *OCEANS 2015 - MTS/IEEE, Washington* (2015)
7. R. Vermaak, M.J. Kamper, Design aspects of a novel topology air-cored permanent magnet linear generator for direct drive wave energy converters. IEEE Trans. Ind. Electron. **59**(5), 2104–2115 (2012)
8. T.M. Hayslett, T. Orekan, P. Zhang, Underwater wireless power transfer for ocean system applications, in *OCEANS 2016 MTS/IEEE Monterey* (2016)
9. T. Orekan, Z. Zhao, P. Zhang, J. Zhang, S. Zhou, J. Cui, Maximum lifecycle tracking for tidal energy generation system. Electr. Power Compon. Syst. **43**, 8–10 (2015)
10. W. Cummins, The impulse response function and ship motions. David Taylor Model Basin Reports, Tech. Rep. (1962)
11. J. Falnes, M. Perlin, Ocean waves and oscillating systems: linear interactions including wave-energy extraction. Appl. Mech. Rev. **56**, B3 (2003)
12. N. Faedo, Y. Peña-Sanchez, J.V. Ringwood, Finite-order hydrodynamic model determination for wave energy applications using moment-matching. Ocean Eng. **163**, 251–263 (2018)
13. M. Penalba, G. Giorgi, J.V. Ringwood, Mathematical modelling of wave energy converters: a review of nonlinear approaches. Renew. Sustain. Energy Rev. **78**, 1188–1207 (2017)
14. M. Verduzco-Zapata, F. Ocampo Torres, Study of a 6 DOF wave energy converter interacting with regular waves using 3D CFD, in *11th European Wave and Tidal Energy Conference* (2015)
15. W. Cummins, The impulse response function and ship motions. Schiffstechnik **9**, 101–109 (1962)
16. T.M. Lewis, B. Bosma, A. von Jouanne, T.K.A. Brekken, Modeling of a two-body wave energy converter driven by spectral jonswap waves, in *2013 IEEE Energy Conversion Congress and Exposition* (2013), pp. 329–336
17. WAMIT Inc., Wamit user manual, Tech. Rep. 2016 [Online]. http://www.wamit.com/manualupdate/v72_manual.pdf. Accessed 7 Sept. 2018

Chapter 3
Study and Analysis of Underwater Wireless Power Transfer

3.1 Introduction

Wireless power transfer (WPT) has been revitalized largely because of the fast-expanding market of mobile devices and electric vehicles [1–4]. Recently, WPT is found to be critically needed in distributed ocean systems that consume excessive power [5], such as an autonomous underwater vehicle (AUV), underwater sensing nodes, ocean monitoring devices, etc. The traditional way of recharging/replacing these devices manually is time-consuming, resulting in limited operation range and disrupted services.

One potential solution is to build underwater docking station with electrical connectors [6–8], which, however, suffers from high maintenance costs and limits to near shore applications. Here, WPT is investigated as a potent solution to recharge distributed ocean systems. The ongoing research of WPT systems is mainly focused on land-based applications [9]. Frequency tuning that automatically adjusts the operating frequency to maximize system efficiency has been developed in [10–13]. Other methods such as impedance matching [14–16], intermediate field repeaters [17, 18], and the use of meta-materials [19, 20] to achieve higher power transfer efficiency for WPT in air have also been investigated. Nevertheless, transferring power wirelessly in a marine environment, where seawater is a conducting medium, is still an open and challenging problem. Several questions or challenges that need to be addressed are summarized below.

- What are the effects of the highly conductive seawater on a WPT system's electrical parameters?
- What are the effects of seawater on its coil radiation resistance and the losses incurred?
- If the losses are highly frequency-dependent, then how to choose an operating frequency range in which high efficiency power transfer is possible? Besides,

T. Orekan, P. Zhang, *Underwater Wireless Power Transfer*, SpringerBriefs in Energy, https://doi.org/10.1007/978-3-030-02562-5_3

the dynamic nature of the ocean makes the WPT system constantly in motion and leads to time-varying coupling coefficient, which necessitates an optimized design strategy to overcome such expected variability.

Furthermore, control methods used in the WPT system in air, such as impedance matching [14, 16], require information from the transmitter through communication to keep track of changes in the system parameters, making them unviable for an undersea WPT system. Rather, a control method that uses information from only the transmitting or receiving side of the system without a communication network is essential.

To address the above challenges, this book presents a comprehensive UWPT efficiency analysis. The electrical properties of the UWPT coils submerged in seawater are quantified. The frequency-dependent characteristics of the coil impedance, the self-inductances, the mutual inductance, and the effects of coil shapes are analyzed and compared to those of WPT in the air. A methodology for optimized coil design to increase power transfer efficiency is developed based on the analytical results.

3.1.1   Notation

The key mathematical notations in this chapter are summarized in Table 3.1. $\oint\!\!\!\oint$ represent the circular integrals over the path of a single turn of each coil; $|\cdot|$

Table 3.1 Nomenclature adopted in this chapter

Notation	Description
T_x	Transmitter coil on the primary side
R_x	Receiver coil on the secondary side
$k_{p/s}$	Coupling coefficient
L_p	Self-inductance on the primary side circuit
L_s	Self-inductance on the secondary side circuit
L_m	Mutual inductance between primary and secondary side
C_p	Series-resonance capacitor on the primary side circuit
C_s	Series-resonance capacitor on the secondary side circuit
R_p	Internal resistances of the primary side coil
R_s	Internal resistances on the secondary side coil
R_{dc}	DC resistance of coil
R_{ac}	AC resistance of coil
R_{rad}	Radiation resistance of coil
R_{rad}^{sea}	Radiation resistance of coil in the seawater
R_{rad}^{air}	Radiation resistance of coil in the air
R_L	Load resistance
V_p	Primary side voltage
V_s	Secondary side voltage
N_p	Number of coil turn on the primary side
N_s	Number of coil turn on the secondary side

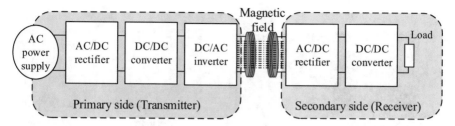

Fig. 3.1 Block diagram of a wireless power transfer system

denotes absolute value of a complex scalar; ω is the frequency in radians; μ is the permeability of medium; σ is the conductivity of the medium [21].

3.1.2 Design

Figure 3.1 shows a typical block diagram of a wireless power transfer system design. The system is made up of two electrical isolated circuits known as the primary side and the secondary side. The primary side consists of the transmitter, and the secondary side consists of the receiver. Transmitter receives low-frequency (50–60 Hz) alternating current (AC) energy source. Since the frequency is too low for induction, the AC is converted to a regulated direct current (DC) voltage using an AC/DC rectifier. In addition, the AC/DC rectifier serves as one of the control unit for regulating the power flow of the entire system by modulating the DC input voltage. The DC voltage, using DC/AC inverter, is then converted to high-frequency AC voltage which is passed through the transmitting coil and generates a magnetic field. The receiver, separated by air gap (mutual inductance), is induced by the magnetic field from the transmitter. Using AC/DC high-frequency rectifier, the voltage on the receiver is adjusted to DC voltage. The voltage is then converted to a suitable DC voltage, depending on the load requirement, using DC/DC buck, boost, or buck-boost converter.

A number of different compensation topology, discussed in Sect. 3.1.2.1, are possible. The compensation of the primary side coil is designed to achieve a resonant frequency equal to the operating frequency of the power source, while the compensation of the secondary side coil is designed to transfer power to the load at a maximum efficiency.

3.1.2.1 Compensation Topology

Basically, there are four compensation topologies as depicted in Fig. 3.2: series-series (SS), series-parallel (SP), parallel-series (PS), and parallel-parallel (PP) topologies. In the SS compensation topology, the resonant capacitor can be designed

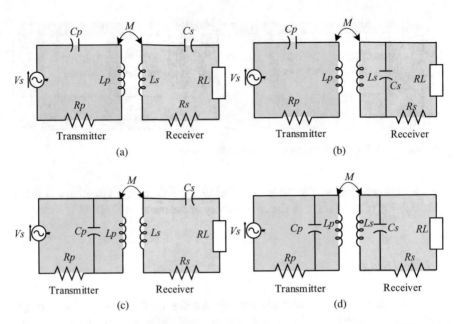

Fig. 3.2 Compensation topology: (**a**) series-series (SS); (**b**) series-parallel (SP); (**c**) parallel-series (PS); (**d**) parallel-parallel (PP)

to resonate with the leakage inductance, which could achieve a higher ratio between the reactive and active power. In [22], an inductive resonant SS topology is used to enhance lateral displacement tolerance. Although, the misalignment tolerance is weak in the case of SS, but the power output capacity is increased compared to PP topology. This is mainly due to the lack of reflected reactant in SS compensated topology, when working at the resonant frequency. For SP compensation topology, the resonant capacitor of the primary side coil can be obtained based on the reflected impedance of the secondary side coil. In contrast to SS compensation topology, the SP compensation topology is more sensitive to the change of resonant frequency, and efficiency decreases when the coupling coefficient decreases. Hence, SP compensation topology is dependent on the coupling effect. In [23], it is shown through comparative analysis that SP compensation topology is more suitable for large loads. Also, a primary SP type compensation network is presented in [24]. A new type of topology, series-series/series-parallel (SS/SP), which can transfer power given the high misalignment between the primary and secondary coils, is presented in [25]. In the PP and PS compensation topologies, the reflected impedance is affected by the load impedance; therefore, the parallel compensation on the primary side circuit is dependent on both the load and the mutual inductance.

3.2 UWPT System Design

Figure 3.3 depicts the equivalent circuit model of the series-series (SS) inductive UWPT system. Accordingly, the SS compensation topology is widely adopted in dynamic inductive WPT, due to its less dependence on the mutual inductance. In addition, it allows high power to flow on the receiver side, and it is proven to be more efficient than series-parallel and parallel-series circuit topologies [26, 27]. The circuit model includes the 3D geometry of the coils in a marine environment. In the primary side of the system is a transmitter coil (Tx) and in the secondary side is the receiver coil (Rx), which are magnetically coupled by a coupling coefficient denoted as $k_{p/s}$. L_p, L_s, C_p, C_s, R_p, and R_s represent inductance, series-resonance capacitors, and internal resistances of the primary and secondary circuit, respectively, and the primary and secondary side voltages are defined as V_p and V_s, respectively. The load resistance is denoted as R_L. From the equivalent circuit of Fig. 3.3, Kirchhoff's voltage law (KVL) equations can be derived as

$$\begin{bmatrix} V_p \\ V_s \end{bmatrix} = \begin{bmatrix} R_p & j\omega_0 k_{p/s}\sqrt{L_p L_s} \\ j\omega_0 k_{p/s}\sqrt{L_p L_s} & R_s \end{bmatrix} \begin{bmatrix} I_p \\ I_s \end{bmatrix} \tag{3.1}$$

From Eq. (3.1), the secondary voltage to primary voltage ratio can be calculated as

$$\frac{V_s}{V_p} = j\frac{(\omega_0 L_m) R_L}{(R_L + R_s)(R_p R_s + (\omega_0 L_m)^2)} \tag{3.2}$$

Subsequently, the secondary current to primary current ratio can be calculated as

$$\frac{I_s}{I_p} = j\frac{\omega_0 L_m}{R_L + R_s} \tag{3.3}$$

In the case, transmitting frequency from the power source is the same as the resonance frequency of the UWPT coils. The power efficiency transfer (η_{pt}) can be written as

$$\eta_{pt} = \frac{(\omega_0 L_m)^2 R_L}{(R_L + R_s)(R_L R_p + R_p R_s + (\omega_0 L_m)^2)} \tag{3.4}$$

where

$$\omega_0 = \frac{1}{\sqrt{L_p C_p}} = \frac{1}{\sqrt{L_s C_s}} \tag{3.5}$$

corresponds to the resonance frequency ($\omega_0 = 2\pi f_0$) and L_m is the mutual inductance which is directly related to the distance between the coils. Here, L_m is responsible for how much power is transferred, and its normalized form, coupling

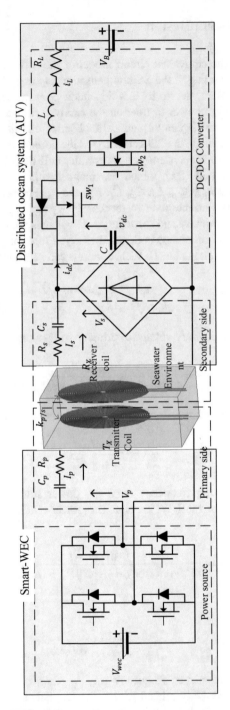

Fig. 3.3 UWPT system configuration

coefficient $k_{p/s}$ (Eq. (3.6)), is used to indicate whether the UWPT system is over-coupled, critically coupled, or under-coupled (cite).

$$k_{p/s} = \frac{L_m}{\sqrt{L_p L_s}}, \qquad 0 \le k_{p/s} \le 1 \tag{3.6}$$

The ocean is a highly electrically conductive medium, which could create several issues for WPT. For instance, the undersea environment could modify the electrical characteristics of the UWPT coils and deteriorate the efficiency of the overall system. In light of this, a critical question is how to optimize the coil design to improve efficiency? The following section investigates the electrical properties of the system coils in seawater environment.

3.3   Coil Analysis in Underwater

In this section, we aim to understand the essential UWPT parameters, such as the coils' resistance, self-inductance, mutual inductance, and resonance frequency, which are critical for optimized coil design and control purposes.

3.3.1   Resistance in Air Compared to Seawater

The total resistance of a coil in air or seawater is the sum of the coil's DC resistance (R_{dc}) which depends on the conductor size, the AC resistance (R_{ac}) which is due to its skin depth, and the radiation resistance (R_{rad}). Since the values of R_{dc} and R_{ac} depends on the coil configuration and the quality factor of the wire, they remain almost equal either in seawater or air. However, owing to the conductive nature of seawater ($\sigma_{seawater} \approx 4\,\text{S/m}$), electric currents are induced by the magnetic field surrounding the coils. Hence, R_{rad} could differ significantly in seawater than in air [28, 29]. Here, we examine the coil immersed in the seawater and compare it to that in the air. The radiation resistance of coil in a conductive medium, derived by Kraichman, is given by [30, 31]

$$R_{rad}^{sea} = \omega\mu a \left[\frac{4}{3}(\beta a)^2 - \frac{\pi}{3}(\beta a)^3 + \frac{2\pi}{15}(\beta a)^5 - \cdots \right] \tag{3.7}$$

And the radiation of coil in free space is given as

$$R_{rad}^{air} = \frac{\pi}{6}\frac{\omega^4 \mu a^4}{c^3} \tag{3.8}$$

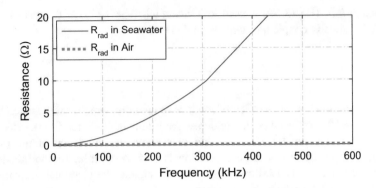

Fig. 3.4 Simulation result showing comparison between radiation resistance in seawater and air (coil radius = 10 cm, number of turns = 20, wire radius = 0.4 mm)

where ω is the frequency, a is the radius of the coil loop in meters, μ is the permeability of the medium, $\beta = (\mu\omega\sigma/2)^{1/2}$, and σ is the medium conductivity. From the expression in Eq. (3.7), we can deduce that the conductivity of the ocean would increase the radiation resistance of the coil. This is in contrast with the expression in Eq. (3.8).

The following conclusions can be observed from the simulation in Fig. 3.4:

- R_{rad} of the coil in the air is minuscule or almost negligible.
- R_{rad} increases exponentially when the frequency is greater than 200 kHz in seawater.

Given these observations, UWPT resonance frequency must operate under some thresholds, e.g., 170–200 kHz. Otherwise, seawater begins to have detrimental effect on the system, which could lead to significant power losses.

3.3.2 Inductance

The self-inductance—the ability of a device or component to store energy in the form of a magnetic field of a circular coil can be calculated using (3.5). When the transmitter and receiver coils are in proximity, the current flowing through one of these devices induced a voltage onto the other. This ability to transfer energy inductively is measured by the mutual inductance [32, 33], as expressed below:

$$L_m = \frac{\mu_0}{4\pi} N_p N_s \oint\!\!\!\oint \frac{e^{-\gamma|R_s-R_p|}}{|R_s - R_p|} dl_p \cdot dl_s \qquad (3.9)$$

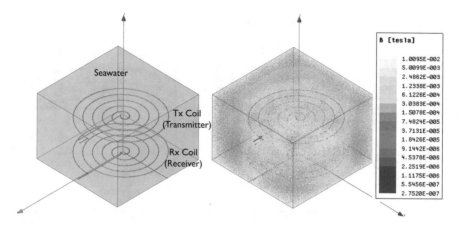

Fig. 3.5 Magnetic flux distribution in UWPT coil structure in seawater

where $\gamma \approx \sqrt{j\omega\mu_0\sigma}$, N_p, and N_s are the number of turns of the primary and secondary coils and the circular integrals are over the path of a single turn of each coil.

Equation (3.9) applies to most coil structures except when the coil has very few numbers of turns and the pitch is very large relative to the wire diameter. From the expression, the imaginary component of the mutual inductance represents loss. Figure 3.5 shows the magnetic flux variation and the influence of seawater on the characteristics of the mutual inductance between the coils. Actually, the magnetic linkages coupling the coils are significantly reduced as the distance between Tx and Rx increases, resulting in fast-dropping mutual inductance (see Fig. 3.6c).

Simulation results have been obtained on the UWPT system with spiral-type coils, as shown in Fig. 3.6a–d. The following conclusions are made:

- The values of self-inductance of Tx and Rx and the mutual inductance slightly increase as the radius of their coil wires increases from 0.4 to 1 mm (see Fig. 3.6a–c)
- The Rx self-inductance varies with distances between Tx and Rx, while the Tx self-inductance remains constant (see Fig. 3.6a, b). Given identical Tx and Rx coil radii, the peak value of the self-inductance of Rx can become higher than that of Tx, resulting in compromised power transfer efficiency (see Fig. 3.6d). The variation in Rx coil is observed to be attributed to the rapid change in distance when the coil is moved away from the Tx coil, in a highly conductive saltwater environment. We suspect that during this change, the Rx coil behaves as if it had a solid conductive core to some degree. However, it can be noticed in Fig. 3.6b that the Rx inductance increase is relatively small and begins to return to its original value as the distance approaches 6 cm.

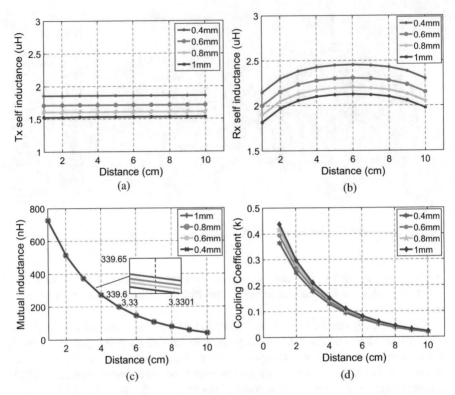

Fig. 3.6 (**a**) Tx self-inductance versus distance at a different wire radius. (**b**) Rx self-inductance versus distance at a different wire radius. (**c**) Mutual inductance versus distance at a different wire radius. (**d**) Coupling coefficient versus distance at a different wire radius

Consequently, to achieve maximum power transfer, the Tx and Rx coil parameters cannot be identical. Rather, the radius of Rx coil should be made smaller than that of Tx.

3.3.3   Coil Shape

It is interesting to determine a UWPT coil shape or topology best suited for integrating ocean distribution system (e.g., AUV). In this chapter, the characteristics of spiral-type coil (refer to Fig. 3.5) and helix-type coil [34] in the undersea environment are comparatively investigated. The design parameters such as coil radius, wire diameter, pitch, and the number of turns of both shapes are made identical for a fair comparison of the effect of the coil shapes. Simulation results are illustrated in Fig. 3.7a–d. It can be seen:

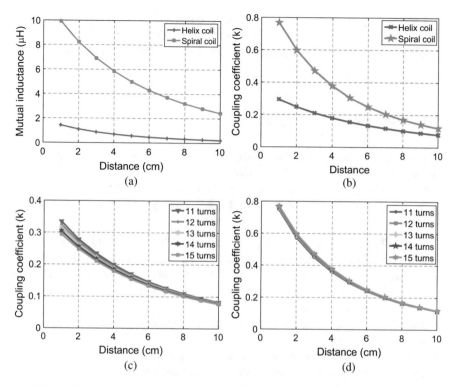

Fig. 3.7 (**a**) Comparison of the mutual inductance between helix and spiral coil shape. (**b**) Comparison of the coupling coefficient between helix and spiral coil shape. (**c**) Coupling coefficient versus distance for helix-type coil. (**d**) Coupling coefficient versus distance for spiral-type coil

- Given the change in distance, the coupling coefficient between the helix coils is much lower than that of the spiral coils. Subsequently, the spiral-type coil is significantly superior to helix-type coil in terms of much higher mutual inductance and as a result higher coupling coefficient. A similar study in [35] shows that when the distance between the coil increases, the efficiency of the spiral coil shape is better than that of helix coil. Therefore, helix-type coil topology is hardly feasible for UWPT application (see Fig. 3.7a, c).
- Increasing the number of turns in helix-type and spiral-type coil topologies, surprisingly, does not improve the coupling coefficient in either case. On the contrary, the coupling coefficient decreases when increasing the number of turns of helix-type coil (see Fig. 3.7c, d).

In real-world situation the wire of the UWPT coil can be protected from seawater using materials such as polyurethane potting and encapsulating materials used on hydrophones. These materials are required to resist the seawater over long periods of submersion. Based on the above analysis of physical and topological characteristics, we can conclude that the following rules should be considered for design and operation of a UWPT system:

1. Low-frequency viability (operate at low frequencies);
2. Differentiation criterion (avoid identical Tx and Rx);
3. Close surface proximity (optimize coil topology to enhance coupling).

3.4 Optimized Coil Design

Enabled by the findings in Sect. 3.3, we present a systematic method for optimized coil design as depicted in Fig. 3.8. The operating frequency should first be determined according to the power level of the UWPT system and frequency-dependent circuit analysis such as that shown in Fig. 3.3. The design procedure is presented as follows:

- *Step 1:* The UWPT system design constraints are imposed based on application, e.g., AUV. The coil shape, relative position of winding, radius of the wire, and number of turns are determined, and the 3D model is constructed in finite element (FE) software. A seawater environment is set up, and the associated Maxwell equations are established.
- *Step 2:* Apply the initial value for the distance between the transmitter coil Tx and the receiver coil Rx.
- *Step 3:* The coil parameters are computed using a 3D FE magnetostatic solver.
- *Step 4:* The values of self-inductances, mutual inductance, and coupling coefficient between the coils are obtained.
- *Step 5:* Compensating capacitors are chosen based on the resonant frequency and the coil inductances using (3.5) and (3.6).
- *Step 6:* Perform parametric sweep on the coils, and obtain optimized coil parameters at various distances.
- *Step 7:* Validate optimized coil design through measurement of the resonant frequency using co-simulation with ANSYS Simplorer. Confirm that the optimal frequency is equal or very close to the operating frequency.
- *Step 8:* The magnetic fields of the coils are calculated based on the optimized parameters of the coil from the magnetostatic analysis.

3.4.1 High-Performance Computing

Since highly precise parameters are needed to get the optimal solution, the above procedure heavily relies on the transient analysis by a FE electromagnetic solver that requires a considerable amount of computational resources. Therefore, high-performance computing (HPC) facility is recommended to solve the large-scale simulations with significantly reduced design cycle. For instance, a simulation of a UWPT with 20-turn coil on a PC could take more than 72 h without guaranteeing

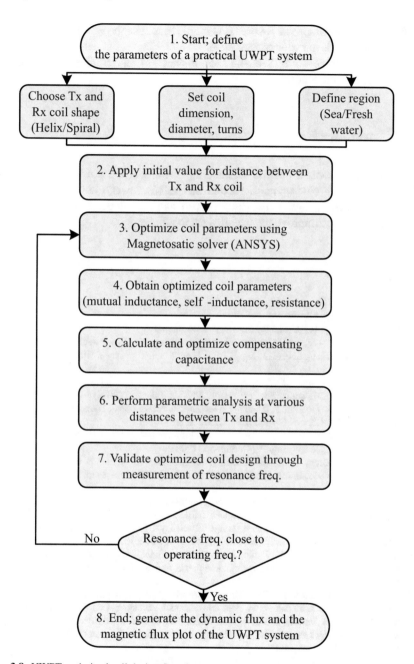

Fig. 3.8 UWPT optimized coil design flowchart

to find a solution, whereas a 8-core HPC is able to reduce the simulation time to less than 12 h and achieve faster convergence.

In Chap. 4, Table 4.1 in Sect. 4.5 is an instance of an optimized UWPT design using the above procedure.

3.5 Summary

This chapter presents a deep understanding of transferring power wirelessly in a dynamic undersea environment. Factors that affect power transfer efficiency are analyzed in detail. A systematic design approach is developed to optimize the coil properties and thus improve the overall system efficiency.

References

1. Witricity [Online], http://www.witricity.com/
2. Qualcomm/haloipt [Online], http://www.qualcomm.com
3. J.M. Miller, O.C. Onar, M. Chinthavali, Primary-side power flow control of wireless power transfer for electric vehicle charging. IEEE J. Emerg. Sel. Top. Power Electron. 3(1), 147–162 (2015)
4. S.Y.R. Hui, W. Zhong, C.K. Lee, A critical review of recent progress in mid-range wireless power transfer. IEEE Trans. Power Electron. 29(9), 4500–4511 (2014)
5. T.M. Hayslett, T. Orekan, P. Zhang, Underwater wireless power transfer for ocean system applications, in OCEANS 2016 MTS/IEEE Monterey (2016)
6. R.S. McEwen, B.W. Hobson, L. McBride, Docking control system for a 54-cm-diameter (21-in) AUV. IEEE J. Ocean. Eng. 33(4), 550–562 (2008)
7. R. Stokey, B. Allen, T. Austin, Enabling technologies for REMUS docking: an integral component of an autonomous ocean-sampling network. IEEE J. Ocean. Eng. 26(4), 487–497 (2001)
8. K. Teo, E. An, P.J. Beaujean, A robust fuzzy autonomous underwater vehicle (AUV) docking approach for unknown current disturbances. IEEE J. Ocean. Eng. 37(2), 143–155 (2012)
9. W. Zhong, S.Y.R. Hui, Maximum energy efficiency tracking for wireless power transfer systems. IEEE Trans. Power Electron. 30(7), 4025–4034 (2015)
10. A.P. Sample, D. Meyer, J.R. Smith, Analysis, experimental results, and range adaptation of magnetically coupled resonators for wireless power transfer. IEEE Trans. Ind. Electron. 58(2), 544–554 (2011)
11. N.Y. Kim, K.Y. Kim, J. Choi, C.W. Kim, Adaptive frequency with power-level tracking system for efficient magnetic resonance wireless power transfer. Electron. Lett. 48(8), 452–454 (2012)
12. B.H. Waters, A.P. Sample, P. Bonde, J.R. Smith, Powering a ventricular assist device (VAD) with the free-range resonant electrical energy delivery (FREE-D) system. Proc. IEEE 100(1), 138–149 (2012)
13. Z. Pantic, K. Lee, S.M. Lukic, Receivers for multifrequency wireless power transfer: design for minimum interference. IEEE J. Emerg. Sel. Top. Power Electron. 3(1), 234–241 (2015)
14. J. Park, Y. Tak, Y. Kim, Y. Kim, S. Nam, Investigation of adaptive impedance matching methods for near-field wireless power transfer. IEEE Trans. Antennas Propag. 59(5), 1769–1773 (2011)

15. L. Huang, A.P. Hu, A.K. Swain, Y. Su, Z-impedance compensation for wireless power transfer based on electric field. IEEE Trans. Power Electron. **31**(11), 7556–7563 (2016)
16. T.C. Beh, T. Imura, M. Kato, Y. Hori, Basic study of improving efficiency of wireless power transfer via magnetic resonance coupling based on impedance matching, in *IEEE International Symposium on Industrial Electronics*, 7 July 2010
17. F. Zhang, S.A. Hackworth, W. Fu, C. Li, Z. Mao, M. Sun, Relay effect of wireless power transfer using strongly coupled magnetic resonances. IEEE Trans. Magn. **47**(5), 1478–1481 (2011)
18. D. Ahn, S. Hong, A study on magnetic field repeater in wireless power transfer. IEEE Trans. Ind. Electron. **60**(1), 360–371, (2013)
19. M.J. Chabalko, J. Besnoff, D.S. Ricketts, Magnetic field enhancement in wireless power with metamaterials and magnetic resonant couplers. IEEE Antennas Wirel. Propag. Lett. **15**, 452–455 (2015)
20. E.S. Rodríguez, A.K. RamRakhyani, D. Schurig, Compact low-frequency metamaterial design for wireless power transfer efficiency enhancement. IEEE Trans. Microw. Theory Tech. **64**(5), 1644–1654 (2016)
21. T. Orekan, P. Zhang, C. Shih, Analysis, design and maximum power efficiency tracking for undersea wireless power transfer. IEEE J. Emerg. Sel. Top. Power Electron. **6**(2), 843–854 (2017)
22. J. Huh, S.W. Lee, W.Y. Lee, G.H. Cho, C.T. Rim, Narrow-width inductive power transfer system for online electrical vehicles. IEEE Trans. Power Electron. **26**(12), 3666–3679 (2011)
23. C. Fang, J. Song, L. Lin, Y. Wang, Practical considerations of series-series and series-parallel compensation topologies in wireless power transfer system application, in *2017 IEEE PELS Workshop on Emerging Technologies: Wireless Power Transfer (WoW)* (2017), pp. 255–259
24. S. Wang, J. Chen, Z. Hu, C. Rong, M. Liu, Optimisation design for series-series dynamic WPT system maintaining stable transfer power. IET Power Electron. **10**(9), 987–995 (2017)
25. Y. Wang, Y. Yao, X. Liu, D. Xu, L. Cai, An LC/S compensation topology and coil design technique for wireless power transfer. IEEE Trans. Power Electron. **33**(3), 2007–2025 (2018)
26. M. Ishihara, K. Umetani, H. Umegami, E.Hiraki, M. Yamamoto, Quasi-duality between SS and SP topologies of basic electric-field coupling wireless power transfer system. Electron. Lett. **52**(25), 2057–2059 (2016)
27. T. Campi, S. Cruciani, F. Maradei, M. Feliziani, Near-field reduction in a wireless power transfer system using LCC compensation. IEEE Trans. Electromagn. Compat. **59**(2), 686–694 (2017)
28. K. Iizuka, R. King, C. Harrison, Self- and mutual admittances of two identical circular loop antennas in a conducting medium and in air. IEEE Trans. Antennas Propag. **14**(4), 440–450 (1966)
29. A. Jenkins, V. Bana, G. Anderson, Impedance of a coil in seawater, in *IEEE Antennas and Propagation Society International Symposium (APSURSI)* (2014)
30. M.B. Kraichman, Impedance of a circular loop antenna in a infinite conducting medium. J. Res. Natl. Bur. Stand. Radio Propag. **66D**(4), 499–503 (1962)
31. J.R. Wait, Insulated loop antenna immersed in a conducting medium. J. Res. Natl. Bur. Stand. **59**(2), 133–137 (1957)
32. S. Babic, F. Sirois, C. Akyel, C. Girardi, Mutual inductance calculation between circular filaments arbitrarily positioned in space: alternative to grover's formula. IEEE Trans. Magn. **46**(9), 3591–3600 (2010)
33. C. Zhang, W. Zhong, X. Liu, S.Y.R. Hui, A fast method for generating time-varying magnetic field patterns of mid-range wireless power transfer systems. IEEE Trans. Power Electron. **30**(3), 1513–1520 (2015)
34. P. Hadadtehrani, P. Kamalinejad, R. Molavi, S. Mirabbasi, On the use of conical helix inductors in wireless power transfer systems, in *IEEE Canadian Conference on Electrical and Computer Engineering (CCECE)* (2016)
35. X. Shi, C. Qi, M. Qu, S. Ye, G. Wang, L. Sun, Z. Yu, Effects of coil shapes on wireless power transfer via magnetic resonance coupling. J. Electromagn. Waves Appl. **28**(11), 1316–1324 (2014)

Chapter 4
Maximum Power Efficiency Tracking for UWPT

4.1 Introduction

The marine environment is highly uncertain, and various external forces will cause disturbances on the UWPT system. The coupling coefficient $k_{p/s}$ that determines the power efficiency will fluctuate due to the motion between Tx and Rx (e.g., wave energy converter station and AUV). Here a novel maximum power efficiency tracking (MPET) that maintains the highest possible power transfer efficiency is presented [1]. For a UWPT deployed in the marine environment, the power efficiency is dynamically changing with the load conditions and the coupling coefficient between the coils. There is a need to track the operating point corresponding to the maximum power efficiency. In solar photovoltaic or wind power technology, maximum power point tracking (MPPT) is a well-known concept. In UWPT system, however, maximum power efficiency is the most important property to be maintained, mainly due to the requirement to reduce power losses and preserve energy which is the most scarce resource for undersea applications. Another major difference from traditional MPPT is that the presented MPET does not need any wireless communication between the transmitter and the receiver which is highly desirable for smart ocean systems. In fact, the k-nearest-neighbor-based machine learning approach is used to estimate the coupling coefficient, making the MPET attractive in practical applications where communication is not viable.

4.2 MPET Design

First, Eq. (3.2) is reexamined to get the secondary voltage:

$$V_s = \frac{(\omega_0 L_m) R_L}{(R_L + R_s)(R_p R_s + (\omega_0 L_m)^2)} V_p \qquad (4.1)$$

© The Author(s), under exclusive licence to Springer Nature Switzerland AG 2019
T. Orekan, P. Zhang, *Underwater Wireless Power Transfer*, SpringerBriefs in Energy,
https://doi.org/10.1007/978-3-030-02562-5_4

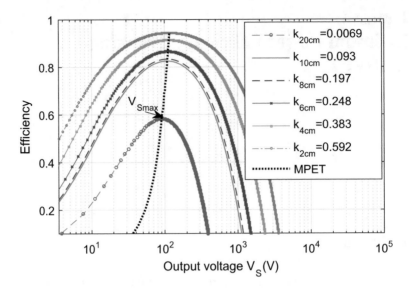

Fig. 4.1 Output voltage characteristics

V_s changes when the distance between Tx and Rx varies. Figure 4.1 depicts the relationship between the V_s and power transfer efficiency at different $k_{p/s}$, under the assumption that the amplitude of the primary voltage V_p and the resistance of the primary side coil are constant. As can be seen, to achieve maximum power efficiency at any specific $k_{p/s}$, V_s should be adjusted to an optimal value $V_{s_{max}}$. The reason why $V_{s_{max}}$ (Eq. (4.2)) leads to maximum power efficiency is because $V_{s_{max}}$ corresponds to a desirable R_L satisfying the impedance matching condition.

$$V_{s_{max}} = \sqrt{\frac{R_s}{R_p}} \frac{\omega_0 k_{p/s} \sqrt{L_p L_s}}{\sqrt{R_p R_s} + \sqrt{R_p R_s + (\omega_0 k_{p/s})^2 L_p L_s}} V_p \qquad (4.2)$$

Finding $V_{s_{max}}$ becomes a significant challenge as it relies on the real-time information of $k_{p/s}$. One approach is to use in-band wireless communication to transmit information between Tx and Rx, in order to achieve impedance matching. Such approach, however, is no longer viable in a marine environment since fast and cost-effective underwater communication is currently unavailable [2]. To tackle this challenge, we present a novel MPET method to force $V_s \to V_{s_{max}}$ in real-time using information from the Tx side only without relying on wireless communication.

Figure 4.2 summarizes our MPET approach. To be specific, a k-nearest neighbors (kNN) is introduced to estimate $k_{p/s}$ using only Rx information (V_s, I_s). Subsequently, $V_{s_{max}}$ is calculated using (4.2) and serves as the reference for feedback control to track. As a salient feature, our MPET is an adaptive tracker that updates the PI controller in real time, and it employs a DC/DC converter to achieve the

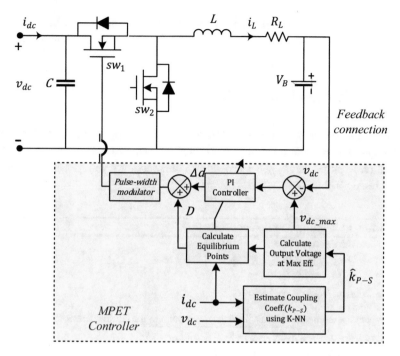

Fig. 4.2 MPET control block

task of changing output voltage to desired voltage by adjusting the duty ratio. This feature makes the UWPT more resilient to the fast-changing $k_{p/s}$.

4.3   Coupling Coefficient Between Tx and Rx

4.3.1   Estimation of Coupling Coefficient

Kirchhoff's voltage law (KVL) equations for the UWPT system in Fig. 3.3 (Tx and Rx sides) can be derived as

$$
\begin{aligned}
V_p &= R_p I_p + \omega_0 k_{p/s} I_p \sqrt{L_p L_s} \\
V_s &= \omega_0 k_{p/s} \sqrt{L_p L_s} I_s + R_s I_s
\end{aligned}
\tag{4.3}
$$

Thus $k_{p/s}$ can be estimated by

$$
\hat{k}_{p/s} = \frac{V_p \pm \sqrt{V_p^2 - 4 R_p I_s (V_s + R_p I_s)}}{2 I_s \omega_0 \sqrt{L_p L_s}}
\tag{4.4}
$$

$\hat{k}_{p/s}$ is subject to uncertainties on the Tx and Rx sides. Furthermore, it is infeasible to track both Tx and Rx states in real time without communication.

4.3.2 Estimation Using KNN

To tackle the challenges, we proposed a data-driven approach, kNN, for robust estimation of $\hat{k}_{p/s}$ under uncertainties, through local approximation using recursive procedure. In particular, $\hat{k}_{p/s}$ is learned from the secondary voltages and currents as shown in Algorithm 1.

Algorithm 1 Coupling coefficient estimation

1: Generate the dataset of pairs:
   $(x_1, y_1), (x_2, y_2), ..., (x_N, y_N)$ where,
   $$y_i = V_p + \sqrt{V_p^2 - 4 R_p I_s (V_{s,i} + R_s I_{s,i})}$$
   $$x_i = 2 I_{s,i}\, \omega_0 \sqrt{L_p L_s}$$
2: Set a query point: x_q
3: Compute the Euclidean distance from the query point to the labeled dataset using
   $$distance(x_i, x_q) = \sqrt{a_1(x_i - x_q)^2 + \cdots + a_n(x_n - x_q)^2}$$
4: Sort the labeled samples in ascending order of distance
5: Find k closest x_i in dataset $(x_{NN1}, ... x_{NNk})$ such that for any x_i not in nearest neighbor set
   $distance(x_i, x_q) \geq distance(x_{NNK}, x_q)$
6: Find a heuristic number k of nearest neighbors
7: Estimate the coupling coefficient which $k_{p/s}$ using k-nearest neighbors $\bar{y} = \frac{1}{k}(y_{NN1} + \cdots + y_{NNK})$

Note that the real parts of the dominating poles in the transfer functions from the primary and secondary voltages to the secondary current are much faster than the fluctuation frequency of coupling coefficient [3]. This means that the transient response attenuates quickly and will hold true under dynamic environment. Thus $k_{p/s}$ remains static in the UWPT. This guarantees the validity and robustness of the proposed kNN estimation method.

4.3.3 Estimation Results

This case performs the kNN-based online estimation of the coupling coefficient. For verification purpose, the exact values of $k_{p/s}$ between the transmitter coil and the receiver coil are analytically obtained from the simulated dataset of primary and secondary circuit parameters. The kNN-based online estimation is then performed under a specific condition (e.g., a resonant frequency of 178 kHz and a load impedance of 50 Ω). The comparison between the estimated $\hat{k}_{p/s}$ values and the exact values demonstrates the accuracy and efficiency of the proposed estimation approach (see Fig. 4.3). In fact, it takes less than 6.123 s CPU time to obtain 1000 estimated $\hat{k}_{p/s}$ values with a maximum error of around 5%. It can be seen that the maximum error occurs due to radiation loss when the distance between Tx and Rx increases. This error can be reduced by sampling more data at a longer period and filtering measurement error. This proves the kNN-based approach is well suited for the real-time MPET.

More estimation results under different loading conditions could be obtained by using impedance sweep, to consistently verified the accuracy and effectiveness of the kNN-based estimation approach.

4.4 Converter Design

4.4.1 DC/DC Converter

When reaching steady state, the DC/DC converter in Fig. 4.2 can be represented by an average state-space (see Appendix A.1 for detailed state-space equations) model as follows:

$$0 = \mathbf{A}_{av}X + \mathbf{B}_{av}U$$
$$V_{dc} = \mathbf{C}_{av}X$$

(4.5)

where

$$\mathbf{A}_{av} = \begin{bmatrix} 0 & -\frac{D}{C} \\ \frac{D}{L} & -\frac{R_b}{L} \end{bmatrix}, \mathbf{B}_{av} = \begin{bmatrix} \frac{1}{C} & 0 \\ 0 & -\frac{1}{L} \end{bmatrix}, \mathbf{C}_{av} = \begin{bmatrix} 1 & 0 \end{bmatrix}$$

$$X = \begin{bmatrix} V_{dc} \\ I_L \end{bmatrix}, U = \begin{bmatrix} I_{dc} \\ V_b \end{bmatrix}$$

L, R_b, v_{dc}, and C are the inductance, the battery resistance, the DC-link voltage, and the filter capacitor, respectively. The battery voltage is V_b and the current flowing into the battery is I_L. The DC/DC converter states can then be found, as follows:

$$I_L = \frac{I_{dc}}{D}$$

(4.6)

$$V_{dc} = \frac{V_b D + I_{dc} R_b}{D^2}$$

(4.7)

where

$$D = \frac{V_B \pm \sqrt{V_B^2 + 4V_{dc}I_{dc}R_B}}{2V_{dc}}$$

(4.8)

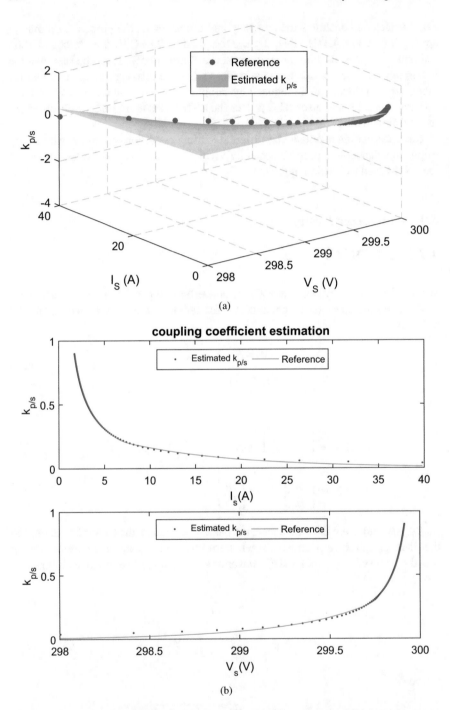

Fig. 4.3 Simulation result showing the comparison between reference coupling coefficient (kp/s) and estimated coupling coefficient ($\hat{k_{p/s}}$) (**a**) Coupling coefficient estimation in 3D (**b**) Coupling coefficient estimation in 2D

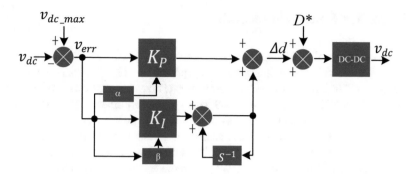

Fig. 4.4 PI controller design

I_{dc}, I_L, V_{dc}, and D are the steady-state values of $i_{dc}(t)$, $i_L(t)$, $v_{dc}(t)$, and Δd, respectively.

During transient state, to maintain a desired level of performance of the MPET, discussed in the next section is a PI controller that adjusts the error caused by the difference between the output voltage V_{dc} and the reference voltage $V_{dc_{max}}$, toward its original steady-state values.

4.4.2 Controller

The control objective here is to adjust the output voltage of the DC/DC buck converter toward the designated output voltage on the load side, under all possible operating conditions. The output voltage of DC/DC converter is a function of the duty cycle Δd. In order to regulate the output voltage, the DC/DC converter should be controlled to regulate its duty cycle.

We introduce an adaptive PI controller (see Fig. 4.4) that adjust PI parameters in real-time [4]. This method is preferable in that it achieves a desired level of performance of the MPET control system when Δd change with time. The proportional constant (K_P) and integral constant (K_I) of the PI are adaptively adjusted by adjusting the values of α and β, respectively. The transfer function of the PI controller is described as follows:

$$G_{PI}(s) = K_P + \frac{K_I}{s} \tag{4.9}$$

α and β (whose values are K_P and K_I, respectively, during the steady state) are adaptively varied as a function of error signal v_{err} value, to drive the steady-state error to zero. As a result, the dynamic output voltage deviation and the settling time of the output voltage are reduced.

4.5   MPET Simulation Results

The effectiveness of MPET has been verified on a UWPT system with a primary voltage V_p of 150 V, a coil operating frequency of 178 kHz, and a switching frequency of 100 kHz in its DC/DC converter. Three test cases are performed by using MATLAB/Simulink and ANSYS Maxwell and Simplorer. The first test case is to examine the resonant frequency at which high power transfer efficiency of UWPT is expected. The second test case is to validate the performance of kNN-based coupling coefficient estimation. Third test case is to verify the proposed MPET approach. The optimized UWPT parameters used in the case studies are listed in Table 4.1.

4.5.1   UWPT Resonance Frequency Analysis

To figure out the resonant frequency, the load of UWPT system is varied from 10 to 100 Ω with a step of 10 Ω, while the distance between Tx and Rx coils is fixed at 1 cm (i.e., $k_{p/s}$ fixed at 0.18).

As shown in Fig. 4.5, there exists an optimal resonance frequency (i.e., 178 kHz) where a maximum efficiency is consistently achieved even though the load varies between 10 and 100 Ω. Therefore, the test UWPT system is to be operated at 178 kHz in the following tests. It should be pointed out, however, that the power transfer efficiency will vary significantly as coupling coefficient between the transmitter coil and the receiver coil varies (see Fig. 4.6). This means that both the online coupling coefficient estimation and MPET are indispensable to maintain high UWPT efficiency in the ever-changing marine environment.

Table 4.1 UWPT system parameters

Parameters	Tx side	Rx side
Coil radius (cm)	6	5
Number of turns	20	18
Wire diameter (mm)	0.6	1
Self-inductance (μH)	47.8	20.3
Parasitic resistance (ohms)	1.3	1.3
Capacitor (nF)	53.1	38.5
Operating frequency (kHz)	178	178
AC voltage source (V)	150	–
Switching frequency (KHz)	–	100
Load resistance (Ω)	–	50

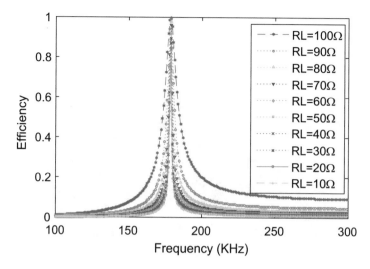

Fig. 4.5 UWPT system efficiency (at $k_{P/S} = 0.18$) as a function of frequency. Target resonance frequency set at 178 kHz for a wide range of load resistance

4.5.2 UWPT Efficiency Maximization

The effectiveness of MPET is extensively evaluated on a dynamic UWPT model subject to fast position changes, by using ANSYS Maxwell and Simplorer and MATLAB/Simulink. The former is employed to perform the transient analysis of the UWPT, in order to obtain the values of the optimized coil parameters. The latter is used to run simulations with the Tustin solver at a time-step of 10 μs to verify MPET.

In this test, the transmitter coil is fixed, while the receiver coil is originally placed at 1 cm and then quickly moved further away at 4, 8, and 10 cm. The output powers and efficiencies of the UWPT system with and without MPET are measured and illustrated in Fig. 4.7. With MPET, when the distance increases, a maximum power efficiency of 85% can be consistently tracked, and a power transfer of 34 W is achieved. Without MPET, however, the output powers decline, and the efficiencies decrease drastically from 85% to 39% as the distance increases. It can also be seen that, when the distance between the coils has a step (or near-step) change, MPET is always able to quickly restore the maximum power efficiency within 0.1 s, which validates the effectiveness and robustness of the MPET.

To further verify the robustness of MPET, the same test is performed in a noisy environment when the output voltage feedback signal and the PI controller are subject to an additive white Gaussian noise with a signal-to-noise ratio of 6 db. Figure 4.8 shows that MPET remains highly efficient and robust under such noisy environment.

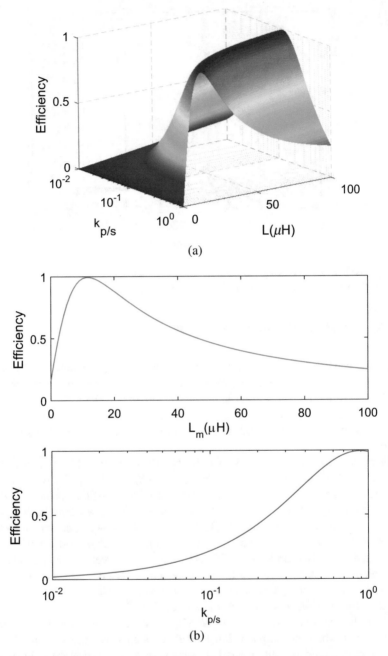

(a)

(b)

Fig. 4.6 Simulation result showing the relationship between the coupling coefficient $k_{p/s}$, mutual inductance L_m, and efficiency at the same resonant frequency $f_0 = 178$ kHz (**a**) 3d diagram (**b**) 2d diagram

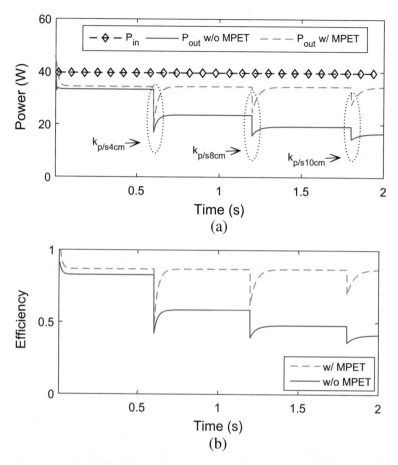

Fig. 4.7 Simulation result of MPET control for UWPT system. (**a**) UWPT power output with and without MPET control. (**b**) Maximum efficiency with and without MPET control

4.6 Experimental Verification

To verify the analysis above, an experimental UWPT prototype is implemented, as shown in Fig. 4.9. The experimental setup consists of a power supply, transmitter and receiver coils, microcontroller and power electronics converter, and a digital oscilloscope. The target resonant frequency is set as 178 kHz, and the parameters used in this experiment system are set to be the same as those in the simulations, except for the input voltage which is set at 30 V due to hardware limitation. All submerged tests for the experiment are implemented in a 20-gallon fish tank filled with seawater of 3.63% salt. Hence, these solutions reasonably model a marine environment. Based on the coil analysis Chap. 3 in Sect. 3.3, the coils are fabricated in a flat spiral disk using 22 AWG wire, and their self-inductances are measured

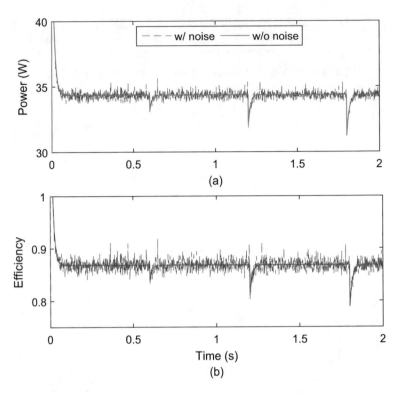

Fig. 4.8 Simulation result showing the response of MPET when subject to additive white Gaussian noise (SNR = 6 db). (**a**) UWPT power output with and without noise. (**b**) Maximum efficiency with and without noise

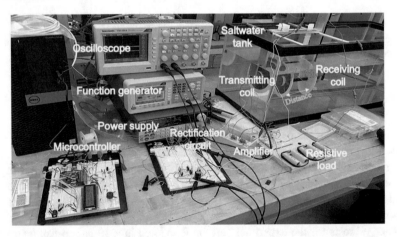

Fig. 4.9 Experimental setup of UWPT

Fig. 4.10 UWPT transmitter and receiver test coil

using a complex impedance analyzer. As shown in Fig. 4.10, the coils are placed onto a 3D printed disks to hold them in place. The input and output cables are connected to the Tx and Rx coil, respectively, and mounted on a test rig that adjusts the distance between the coils.

On the transmitting side of the UWPT system, a DC source is used to provide a constant voltage to an operational power amplifier (APEX PA 94a) at 178 kHz generated by a function generator. The function generator used here could either be replaced by a stand-alone function generator IC or be created using software on a microcontroller. The DC power supply of the power amplifier is set to its maximum range ±30 V, creating an AC voltage of 30 Vpp across the TX coil.

Meanwhile, on the receiving end of the UWPT system, four SBR10U40CT diodes are used to build a standard full-bridge rectifier which converts the transferred AC voltage signal to DC. A DC/DC buck converter described in Sect. 4.5 subsequently used to step down the voltage to a desired voltage level. The control unit ATmega328p includes a 10-bit analog-to-digital converter, which converts the voltage and current to digitized data used to execute MPET.

Figure 4.11 shows the measured results of the UWPT system, when the Rx coil is placed at 1, 2, 3, and 4 cm away from the Tx coil. It can be seen that the voltage received by the RX coil gradually reduces as the distance between the transmitter and the receiver increases. Figure 4.11a–d corresponds to power transfer efficiencies of 80%, 73%, 66%, and 58%, respectively, through seawater. Figure 4.12 gives the measured power efficiency compared against the simulation results, which shows close agreement between simulations and experiments.

Figure 4.13a shows the resonant frequency obtained by applying the fast Fourier transform (FFT) on the raw data from the oscilloscope. The result shows the power transfer efficiency is maximized at the target resonant frequency. As the distance increases from 1 to 10 cm (target distance range of the UWPT), the efficiency is still greater than 80%. However, when the distance increases beyond the target range, the seawater begins to have detrimental effect on the system, and the efficiency

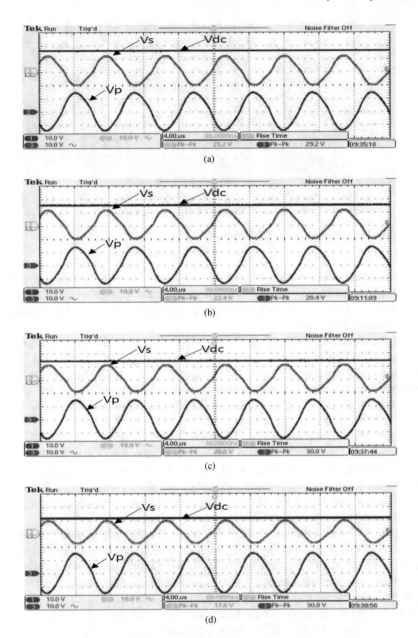

Fig. 4.11 Part of the experimental results. Transmitter and receiver voltage waveform. (**a**) At a distance of 1 cm. (**b**) At a distance of 2 cm. (**c**) At a distance of 3 cm. (**d**) At a distance of 4 cm

quickly decreases below 50% at the resonant frequency of 178 kHz. Figure 4.13b shows the sensitivity of the UWPT system when it deviates from the resonant frequency. Even though a high efficiency of 80% is still feasible when the resonant frequency deviates into a frequency range of [165–185 kHz], a variable capacitor is

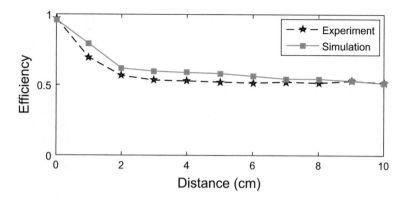

Fig. 4.12 Measured experimental result of the UWPT power transfer efficiency at a distance between 1 and 10 cm

recommended to be used at the transmitter side to bring the resonant frequency to its precise value as in [5].

In Fig. 4.14a, when the distance between the Tx coil and the Rx coil increases without control, the output voltage after rectification drops significantly. In practice, as shown in Fig. 4.14b, the buck converter can be controlled to regulate the output voltage at 12 V in spite of the changes of the distance between Tx and Rx.

4.6.1 Limitation

One limitation of UWPT coils is the sensitivity of misalignment of the coils. Although, stagger-tuning approach has been employed is some applications, but due to variations in the underwater environment, the UWPT will still suffer this problem. This effect of angular or lateral misalignment of the coils could result in decrements in output voltage. Therefore, the presented MPET in this chapter should be improved with an automatic search for the optimum frequency of operation to account for misalignment between the primary and secondary coils while maintaining a high power transfer efficiency. Also, single primary side and multiple secondary sides will necessitate energy management; hence, optimization and coordination of the systems will be investigated in future work.

4.7 Summary

In this chapter, a novel maximum power efficiency tracking (MPET) control is developed to estimate UWPT coupling coefficient in real time through machine learning, to effectively track the maximum power efficiency. Extensive simulation

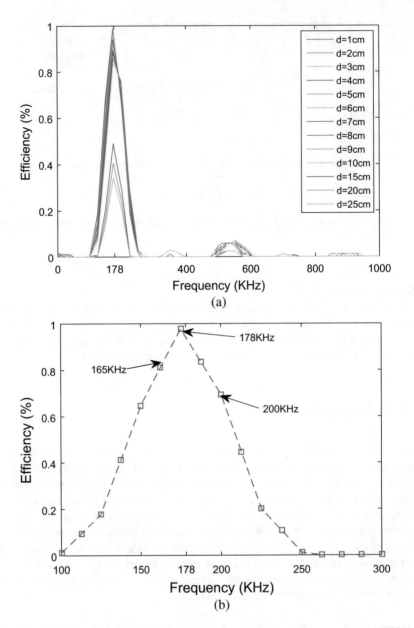

Fig. 4.13 Experimental results of UWPT resonance frequency (target frequency is set at 178 kHz).
(**a**) Resonant frequency at a distance between 1 and 10 cm versus efficiency. (**b**) Sensitivity of the
resonance frequency

and experimental tests validate the effectiveness of the presented method. Future
research will be focused on MPET for multiple-source ocean energy systems and
new power electronic topologies that enable load variation across broader range for
better coupling coefficient estimation.

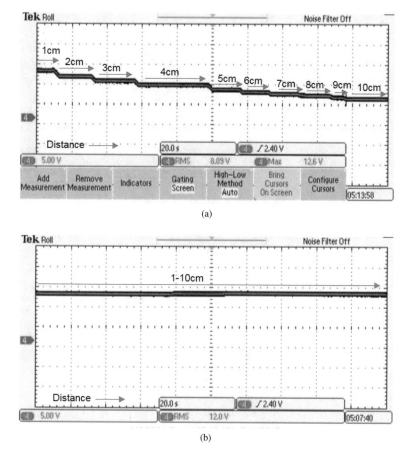

Fig. 4.14 Experimental results of the UWPT system. (**a**) Without control when the distance is adjusted from 1 to 10 cm at unfixed interval. (**b**) Controlled to a regulated output voltage of 12 V

References

1. T. Orekan, P. Zhang, C. Shih, Analysis, design and maximum power efficiency tracking for undersea wireless power transfer. IEEE J. Emerg. Sel. Top. Power Electron. **6**(2), 843–854 (2017)
2. Z. Zeng, S. Fu, H. Zhang, Y. Dong, J. Cheng, A survey of underwater optical wireless communications. IEEE Commun. Surv. Tutorials **19**(1), 204–238 (2017)
3. V. Jiwariyavej, T. Imura, Y. Hori, Coupling coefficients estimation of wireless power transfer system via magnetic resonance coupling using information from either side of the system. IEEE J. Emerg. Sel. Top. Power Electron. **3**(1), 191–200 (2015)
4. V. Arikatla, Adaptive control methods for dc-dc switching power converters, Ph.D. dissertation, University of Alabama, 2011
5. J. Tian, A.P. Hui, A dc-voltage-controlled variable capacitor for stabilizing the ZVS frequency of a resonant converter for wireless power transfer. IEEE Trans. Power Electron. **32**(3), 2312–2318 (2017)

Chapter 5
Energy-Maximizing Control for Ocean Energy Converter

5.1 Introduction

Several approaches for control of tidal current converter and WECs have been performed, in the pursuit of high efficiency or power production. On one hand, in tidal energy conversion system, operation, maintenance, and power production are the major consideration when developing control strategies. On the other hand, in WEC, some control methods are related to the array configuration of the devices, and others analyze them as isolated devices. Although the control strategies vary depending on the type of PTO (e.g., linear generator or hydraulic), they share one primary goal: to maximize the system's power output [1–4]. Among these techniques, passive control strategies are the easiest to implement but also have lower absorption of wave energy [5]. The two most popular control strategies for linear generator type WEC are the reactive control and the latching control. While the reactive control uses machinery force to ensure optimal control for the WEC's motion, a latching control method forcibly locks the WEC's motion when the velocity is close to zero and then releases it after a certain duration or "latching time" [6, 7]. This keeps the velocity in phase with the excitation force so that the maximum amount of power can be extracted. Even though the use of a latching control method would theoretically allow for maximal energy conversion, its practicality has been questioned in [8], due to the additional mechanical configuration required to hold the WEC.

The following are recent control techniques purposed for maximizing power absorption by ocean energy converters.

- Tom and Yeung [9] developed a nonlinear model predictive control (NMPC), which is applied to a model-scale point absorber to maximize its power absorption, in regular and irregular sea-state. The control strategy was focused on determining the optimum PTO damping that could be turned on/off. The result shows that power absorption was 50% greater in NMPC than passive system.

© The Author(s), under exclusive licence to Springer Nature Switzerland AG 2019
T. Orekan, P. Zhang, *Underwater Wireless Power Transfer*, SpringerBriefs in Energy,
https://doi.org/10.1007/978-3-030-02562-5_5

- Sousounis et al. [10] present a variable speed control strategies and a single-tuned filter in order to mitigate characteristic harmonics generated, thereby allowing the system to operate at maximum power coefficient. The results in this work show the controller operating frequencies can be controlled to resonant minimum frequencies and therefore reduce the system harmonics.
- Villa-Jaen et al. [11] proposed using both reactive and latching control strategies. This method is adopted while considering the copper losses reduction in the electric generator and the limit of the maximum excursion constraint, simultaneously. The results, obtained through linear approximations, are compared to conventional latching control. In normal wave period, the performance of the controls is similar but differs at peak wave periods. The proposed control improves the peak-to-average power ratios.

5.2   MPEC for Smart-WEC

In the wave power conversion process, the power transfer efficiency of the device is calculated by comparing the mechanical energy absorbed by the system buoy and the electrical energy generated by the DDLG. A maximum power efficiency control (MPEC) formulated in a discrete time domain is presented. The control's objective is to calculate the power electronic converter voltage needed to force the measured current to its reference value. Due to the reciprocating motion of the WEC which causes the variable frequency and amplitude in the electrical output of the linear generator, an AC/DC power rectifier is considered. Hence, the power extracted is stored in a battery energy storage.

5.2.1   Control Strategy

According to maximum power transfer theorem in circuit theory [12], the maximum coil current required to extract maximum power from the Smart-WEC linear generator is given as

$$i_{max}(t) = \frac{e(t)}{2R} \tag{5.1}$$

where $e(t)$ is the EMF and R is the internal resistance of the generator. In theory (Eq. (5.1)), a phase difference between the DDLG current and the EMF will significantly reduce efficiency. In order to achieve maximum energy transfer from the DDLG, the currents must be in phase with the generator's induced EMF with the peak value of the current proportional to the peak value of the EMF. However, it is not possible to measure the induced EMF on load. Sense coils are generally installed on the system to measure and estimate the EMF as in [13, 14]. Such technique, however, is inefficient and unreliable for a Smart-WEC system that operates in a harsh ocean environment.

The use of a sensor-less coil, an approach similar to the one used to calculate EMF in [15, 16], is adopted in the MPEC to achieve a sensor-less control of the EMF. Figure 5.1 shows the presented MPEC controller for the Smart-WEC. Assuming the current flowing into the filter capacitor C_f is the switching frequency current which is ignored in this case, the rectifier voltage V_{rec} can be calculated as

$$V_{rec}(t) = V_t - L_f \frac{d_i(t)}{d(t)} \tag{5.2}$$

where V_t is the DDLG instantaneous terminal voltage and L_f is the filter inductance, respectively. In the presented MPEC strategy, to predict the future values of the current and voltage, a discrete time model is adopted in the succeeding sampling interval $(k + 1)$, from the measured voltage and currents at the kth sample. Using the Euler approximation

$$\frac{dx}{dt} \approx \frac{x[k+1] - x[k]}{T_s} \tag{5.3}$$

Equation (5.2) can be written as

$$V_{rec}[k] = V_t[k] - L_f(i[k+1] - i[k]) \tag{5.4}$$

where T_s is the switching time. For a given rectifier state, i.e., output voltage and the current error $e[k] = i^*[k] - i[k]$, where i^* is the reference current, the current error $e[k + 1] = 0$ of the next sampling time can be predicted. Applying the calculation on Eq. (5.4), the reference rectifier's voltage V^*_{rec} is then given as

$$V^*_{rec}[k] = V_t[k] - L_f(i^*[k] - i[k]) \tag{5.5}$$

5.2.1.1 Current Error Prediction

As shown in the Smart-WEC DDLG featured in Fig. 3.4, the very low frequency operation of the DDLG renders the synchronous inductance L_s negligible when compared to the internal resistance R_s. In fact, this becomes an important advantage for DDLG use in Smart-WEC as it greatly simplifies the calculation of the $e(t)$, which is given as

$$e[k] = V_t[k] + i[k]R_s \tag{5.6}$$

Therefore, by substituting the above equation into Eq. (5.1), the required reference of the maximum current can be calculated as

$$i^*_{max}(t) = \frac{1}{2R}(V_t[k] + i[k]R_s) \tag{5.7}$$

where $1/2R$ is the current scaling factor (k_{sf}). In practice, the current drawn scaled by a factor $k : 0 < k < 1$ from a coil can be controlled proportional to the $e(t)$ [12]. In the MPEC controller, the current scaling factor is $k_{sf} = 0.065$.

The control references in the above equations maximize the power transferred from the DDLG to the power electronics converter.

5.2.2 Simulation Results

The proposed MPEC is implemented for control verification, as seen in Fig. 5.1, and the DDLG is connected to the rectifier in parallel with an LC filter using the design parameters shown in Table 2.1. Simulations are set up for regular and irregular waves, and the power extraction efficiency in both sea-state is compared.

5.2.2.1 Regular Waves

Shown in Fig. 5.2 are the results of power absorbed, electrical power, and terminal voltage of the DDLG. The average power absorbed and electrical power at the DDLG terminal, as seen in Fig. 5.2a, are 266 W and 170 W, respectively. The following conclusions can be drawn from this test case:

- In Fig. 5.2c, $e(t)$ is out of phase with current without control, and it can be seen from the results that the efficiency of the system without control is calculated to be 64%.
- Figure 5.3 shows the performance of the MPEC when the Smart-WEC is subjected to regular wave conditions. In Fig. 5.3a, the c-phase of the induced EMF $e(t)$, phase current $i(t)$, and the corresponding predicted reference current $i^*(t)$ shows that the MPEC control strategy achieves the goal of keeping the voltage and current in phase. This allows maximum power to be extracted by the generator. Figure 5.3b expands the time base during one stroke, to reveal that the phase current is in phase with the generator-induced EMF calculated from the MPEC control.
- And the predicted current reference calculated in the MPEC controller tracks the DDLG phase current as required.
- The captured power profile with MPEC control is shown in Fig. 5.3c. Compared to Fig. 5.2a, the electrical power at the DDLG terminal, which is transferred to the power electronic rectifier, is calculated as $P_{elec} = 247$ W, with a power efficiency ($\eta = P_{elec}/P_{abs}$) of 93%.

As the test case results demonstrate, the MPEC performs well, and the DDLG is able to increase power extraction from 64% to 93% without considering eddy-current losses.

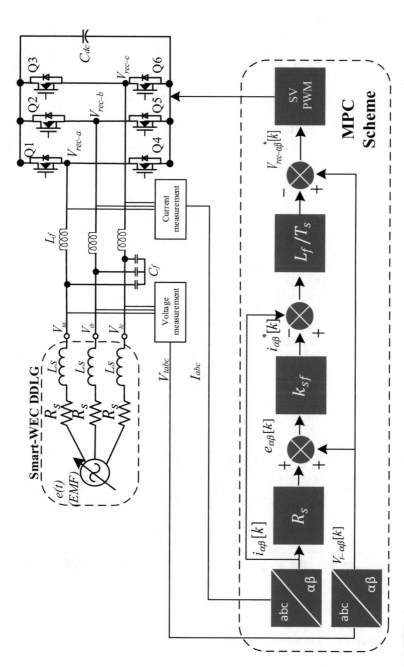

Fig. 5.1 MPEC scheme

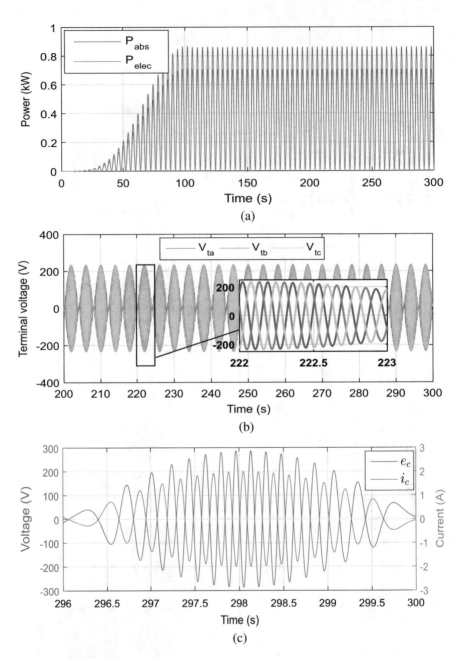

Fig. 5.2 (**a**) Power absorbed by the Smart-WEC. (**b**) DDLG 3-phase terminal voltage. (**c**) Induced EMF voltage out of phase with current

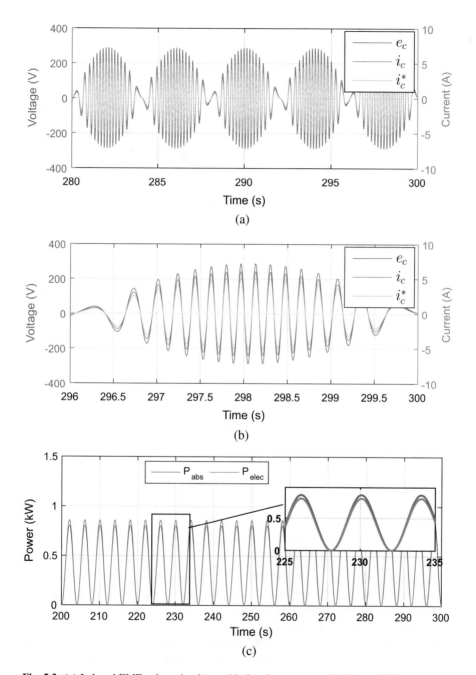

Fig. 5.3 (**a**) Induced EMF voltage in phase with the phase current. (**b**) Induced EMF voltage in phase with the phase current (one translator stroke). (**c**) Power absorbed and electrical power

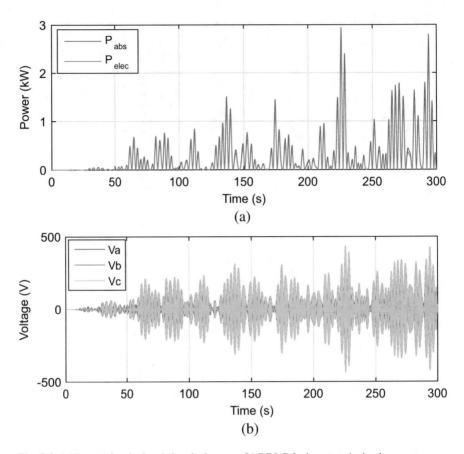

Fig. 5.4 (**a**) Power absorbed and electrical power. (**b**) DDLG 3-phase terminal voltage

5.2.2.2   Irregular Waves

In this test case, compared to the Smart-WEC response when subject to regular waves, the following conclusions can be drawn:

- As seen in Fig. 5.4a, when the device is subjected to realistic irregular wave conditions, it causes large fluctuations in power absorption, indicating that there is a major difference in the power computed using regular waves. The most noticeable difference is the peak-to-average power ratio which exceeds other energy sources, such as wind and solar power. It is evident from these observations that without proper control these slow-power fluctuations may cause significant thermal cycling of the system power electronic converters, which would have to be considerably overrated for such system.
- Figure 5.4b shows the 3-phase terminal voltage of the DDLG. From the results we can see the effect of the reciprocating wave motion on the voltage. Within

each cycle there are zero dips, and a power electronic rectifier is required to enable constant power flow to the load. The average power absorbed is $P_{abs} = 341$ W, almost 75 W more when the system is subject to regular waves. The power at the generator terminal which is transferred to the power electronics, without control as seen in Fig. 5.4a, is $P_{elec} = 211$ W; this results in an efficiency of 62%.

• Figure 5.5 shows the simulation results when the MPEC control is implemented considering irregular wave conditions. Despite the high fluctuation of the induced EMF, the proposed MPEC shows excellent results in terms of forcing the current to track the reference current. In Fig. 5.5a, b, the reference current tracks the phase current well, and it is shown that the induced EMF is in phase with the current. The average power transferred is $P_{elec} = 317$ W, at efficiency of 93%. This resulted in a maximum power extraction shown in Fig. 5.5c.

In the irregular wave conditions, the peak-to-average power ratio and the average power absorbed by the system is higher than that of the regular waves as expected. However, because MPEC increases power transferred to the DDLG in both cases, the maximum power transferred efficiency is almost in agreement: an efficiency increase from 64% to 93%. The fact that the DDLG phase resistance is known and the synchronous inductance is considered negligible aids in the accurate calculation of the DDLG's EMF (and hence the current reference) in the MPEC. Note, however, that the three-phase voltage and current of the generator increase linearly with the Smart-WEC translator speed. Under such conditions the power converters must be able to track the changes in the frequency to extract maximum power. The rating of the linear generator and the power electronics equipment should also be taken into consideration when designing the MPEC controller. Also, for an iron-cored Smart-WEC DDLG, the effect of the higher inductance in iron-cored LG will have to be considered when MPEC is adapted. In addition, the MPEC may not be visible for larger WECs with complex generators, as the nonlinearities of parameters, such as PTO, viscous, and mooring forces, have to be considered.

5.3   MLCT for Tidal Energy Converter

In this section, a maximum life cycle tracking (MLCT) control is presented. As a timely contribution to the emerging field of tidal power technology, MLCT combines the maximum power point tracking (MPPT) and constant torque control strategy to extend the system life cycle [17]. To mitigate stress on the shaft with very little power trade-off, the MLCT methodology applies a constant torque strategy to the tidal turbine at high tidal speed range. Detailed case studies are presented, based on a real-time digital simulator model of a tidal turbine that employs a permanent magnet synchronous generator (PMSG) and connects to the grid through a back-to-back converter interface.

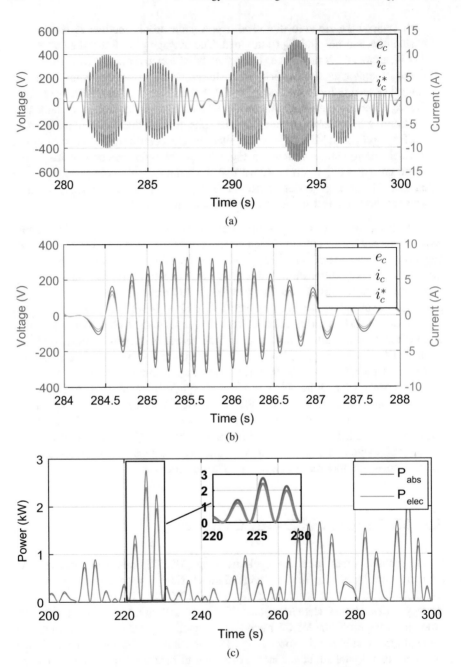

Fig. 5.5 (**a**) Induced EMF voltage in phase with the phase current. (**b**) Induced EMF voltage in phase with the phase current (one translator stroke). (**c**) Power absorbed and electrical power (efficiency increases from 64% to 93%)

5.3.1 Tidal Drivetrain Modeling

The system configuration under study is shown in Fig. 5.6, where the permanent magnet synchronous generator is driven by a tidal turbine. Details of the system model can be found in Appendix B. In the turbine generator system, the mechanical torque is applied to the turbine blades, and the electromagnetic torque is applied to the generator in the opposite direction. The mechanical dynamic equation of a generator is given by

$$J\frac{d\omega}{dt} + D\omega = T_m - T_e \tag{5.8}$$

where T_m is the mechanical torque, T_e is the electromagnetic torque, D is the friction, and J is the joint moment of inertia of the turbine and generator. The single mass equation is adopted because the drivetrain and generator rotates as a whole, and the shaft is quite rigid for modern tidal systems. Moreover, the friction can also be omitted in the analysis without affecting the final results.

The interaction of both mechanical torque and electromagnetic torque causes stress on the turbine blades and shaft, leading to fatigue which eventually results in failure. The relation between torque and the life cycle of the shaft is described through a torque-life curve in Fig. 5.7. The number of cycles can be calculated using the following equation:

$$N = \frac{1}{2}\left(6.4 \times 10^{-6}\frac{\tau}{R_0^3}\right)^{-17.86} \tag{5.9}$$

where N is the number of cycles the shaft can endure, R_0 is the radius of the shaft, and τ is the maximum torque in each cycle. The value (6.4×10^{-6}) is different for every shaft, and it is calculated based on the properties of the shaft material. The equation in (5.9) is used to estimate the life cycle of the shaft under the assumption that the material is the same as that in [18]. Above the fatigue limit in Fig. 5.7, the number of cycles decreases exponentially with stress, indicating the effect of stress on tidal turbine shaft. However, in actual application, the shaft undergoes complex stress variations; therefore the maximum torque is not constant in each cycle. To calculate fatigue damage under these conditions, a standard technique called Miner's rule is used and can be expressed mathematically as

$$D_{life} = \Sigma\frac{n_i}{N_{fi}}, \tag{5.10}$$

where,

n_i = number of cycles at i-th torque,
N_{fi} = number of cycles to failure at i-th torque,
D_{life} = fraction of life expended.

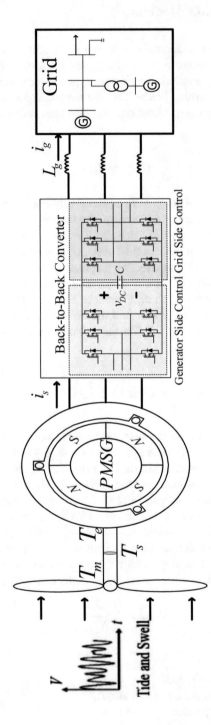

Fig. 5.6 Tidal system configuration

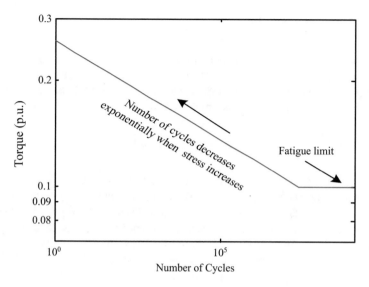

Fig. 5.7 Torque-life curve

When the fraction of life expended equals one, the Miner's rule predicts failure of the shaft. The number of cycles and the magnitude of each cycle are determined by rain-flow counting method [19] (see Appendix B.2).

To analyze the torque on the shaft, we consider the torque at a cross section s in the shaft, which is T_s. The mechanical equations are

$$J_T \frac{d\omega}{dt} = T_m - T_s \tag{5.11}$$

$$J_G \frac{d\omega}{dt} = T_s - T_e \tag{5.12}$$

where J_T is the moment of inertia of the tidal turbine and J_G is the moment of inertia of the generator. The total inertia on the shaft is therefore given by

$$J_T + J_G = J, \tag{5.13}$$

Substitute (5.11) into (5.12), we get

$$T_s = \frac{J_T T_e + J_G T_m}{J} \tag{5.14}$$

The moment of inertia of a tidal generator J_G is generally $10^{-3}\,\text{kg m}^2$, which is quite small compared to the moment of inertia of a tidal turbine which could reach a value as high as $10^4\,\text{kg m}^2$ [20]. Thus, the torque on the shaft T_s is mainly determined by the electromagnetic torque T_e.

5.3.2 Control Strategy

The control strategy for tidal generation system incorporates a maximum power point tracking (MPPT) which adjusts the rotor speed to maximize the turbine output power and a new maximum life cycle tracking (MLCT) strategy. The power harnessed by a tidal turbine can be calculated by the following equation:

$$P = \frac{1}{2}\rho C_P \pi R^2 V^3 \tag{5.15}$$

Here C_P is known as the turbine power coefficient which depends on the turbine blade design and its hydrodynamics. Typically, the optimal C_P value of a tidal turbine for normal operation is in the range of 0.35–0.5 [21]. Figure 5.8 illustrates the tidal turbine characteristics at different tidal speeds, and the extractable power is calculated based on (5.15). In this figure, the black dash line is the conventional MPPT control strategy curve. The mechanical torque T_m is the slope of the straight line connecting the operating point to the origin. It rises fast as the operating point goes up along the maximum power point line. Therefore, when the tidal current speed is very high, e.g., 3 m/s, the mechanical torque T_m increases as well. At the steady state, the electromagnetic torque is equal to the mechanical torque and adds to the stress on the shaft and turbine blades.

To mitigate the stress, a MLCT strategy is introduced as shown in Fig. 5.8. It is a combination of the MPPT control strategy and the constant torque control strategy. At low tidal current speeds, for instance, 1 m/s, the MLCT strategy tracks the maximum power points (MPPs). As the tidal current speed increases to a specified point, it switches and tracks the red or blue line whose extension in the opposite direction passes the origin. Figure 5.8 gives two examples of MLCT strategies.

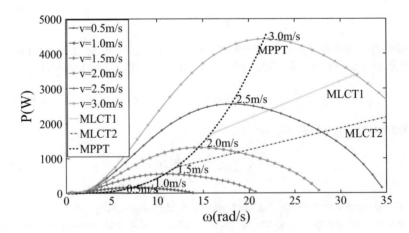

Fig. 5.8 Turbine characteristics and control strategies

The MLCT1 control extracts more power than control MLCT2; however, control 2 applies lower torque on the tidal shaft and blades. Hence, the choice of controller depends on the strength of the shaft.

5.3.2.1 Tidal Speed Estimation and Reference Generation

Traditionally, the measurement of tidal current speed using flowmeter or other types of sensor is necessary for the controllers to generate the reference rotating speed of the turbine. However, these sensors are expensive to install and maintain underwater. Thus, we use artificial neural network (ANN) supplied with captured power and rotating speed to estimate the tidal current speed in real time. Each $P - \omega$ pair in Fig. 5.8 corresponds to a tidal current speed, which shows a nonlinearity behavior with the explicit function $v = f(P, \omega)$. The ANN with several hidden layers of neurons learns this correlation after training and predicts the tidal current speed used in the MLCT controller. Multiple hidden layers are often adopted for better prediction accuracy [22] (see Appendix B.3).

The data generated from each characteristic curve in Fig. 5.8 are obtained to train a feedforward neural network, which is then used to estimate tidal current speed. In Fig. 5.9, the first stage (speed estimation) uses one hidden layer of 60 neurons, and the second stage uses one hidden layer of 5 neurons to generate the reference rotating speed. The data used in training the ANN can be found in Table B.1 in Appendix B. In the table, the first column is the tidal current speed, which serves as the input of the ANN. The second and third columns are the outputs for the MPPT control and the MLCT control. It can be seen from the data that when the tidal current speed is under 2.2 m/s, the inputs of the two neural networks are the same. When it exceeds this speed, the rotating speed under MLCT control becomes higher. Table 5.1 shows the speed estimation samples.

Training the neural networks is done with the MATLAB function "fignet." One can also use "gensim" to convert the neural network function to a Simulink block. This enables the neural networks to be embedded into the real-time simulation environment.

Fig. 5.9 Feedforward neural network

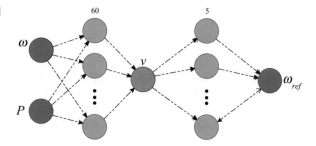

Table 5.1 Data for speed
estimation stage

λ	ω	P
0.9	4.5	436.5
1.8	9.0	1487
2.7	13.5	2291
3.6	18.0	2561
4.5	22.5	2369
5.4	27.0	1881
6.3	31.5	1075

5.3.2.2   Back-to-Back Converter Implementation

Both MLCT and MPPT have been implemented via a back-to-back power electronic
converter system. Since the output power of the tidal turbine fluctuates and
the frequency and the magnitude of the terminal voltage vary, the back-to-back
converters are employed to interface tidal generator system with the power grid
because of its flexible control [23]. It consists of a machine-side converter and a
grid-side converter which connects to the grid through a filter. Between the two
converters is the DC link which absorbs the instantaneous active power difference
and works as a voltage source for the converters. The grid-side converter maintains
a constant DC link voltage and controls the active power and the reactive power
delivered to the grid. The machine-side converter controls power generation, which
is able to apply different control strategies. Control algorithms for the two converters
are decoupled and operate at different AC frequencies. The control block diagram
of the machine-side converter is shown in Fig. 5.10a. Here i_{ds-ref} is the control
reference of the d axis current, and $i_{ds-ref} = 0$; i_{qs-ref} is the control reference of
the q axis current, which plays an important role in extracting the maximum power
from the tides. Zero d-axis current control is applied to ensure a linear relation
between electromagnetic torque and q-axis current for both salient and non-salient
rotor PMSG [24]. Subsequently, the grid-side converter controller, as shown in
Fig. 5.10b, is controlled to output constant active power to the grid and keep the
DC link voltage constant. The q-axis reference current is set to zero to eliminate
reactive power, and all the grid voltage is in the d-axis.

5.3.3   Case Study

In this section, extensive numerical tests have been conducted to validate the
effectiveness of MLCT. In case 1, we use tidal current speed data to test the system
performance under normal conditions. The purpose is to show how MLCT reduces
the mechanical torque and electromagnetic torque in the tidal system and how it
compares to MPPT. In case 2, neural network and sensor speed are used to generate
the reference rotating speed of the turbine to show how a neural network improves
reliability of the tidal generation system.

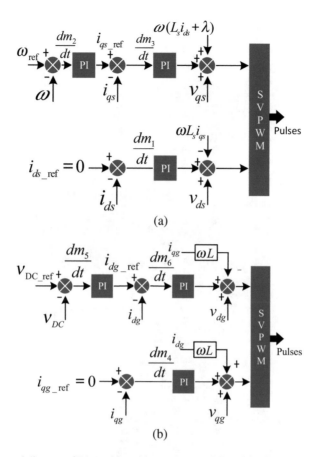

Fig. 5.10 Control diagram of (**a**) machine-side converter and (**b**) grid-side converter

A 12-h simulation is carried-out with massive real tidal current speed data as input. Models with this much detailed and big data can hardly be simulated in conventional PC-based power system simulators such as PSCAD or SimPowerSystems. Hence, to enable fast simulations, Opal-RT real-time hardware-in-the-loop simulator is employed. With multicore processors and large memory, the Opal-RT is able to run simulations in real time and store large amounts of data [25]. A schematic of the model as implemented in the Opal-RT environment is shown in Fig. 5.11.

5.3.3.1 Case I: Normal Condition with Real Tidal Current Data

The MPPT and MLCT performance are compared. Simulation results in Fig. 5.12 shows that at low tidal current speeds, both strategies capture the maximum power available. However, when the tidal current increases, the MPPT captures the most

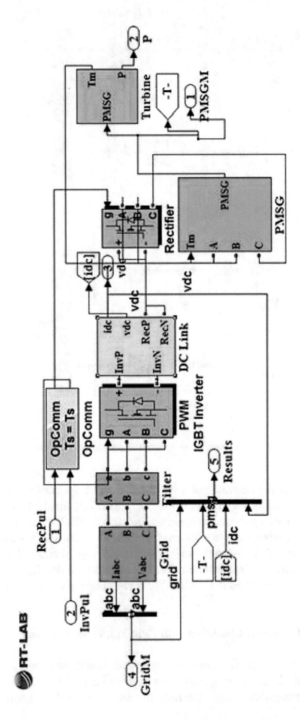

Fig. 5.11 RT-lab model

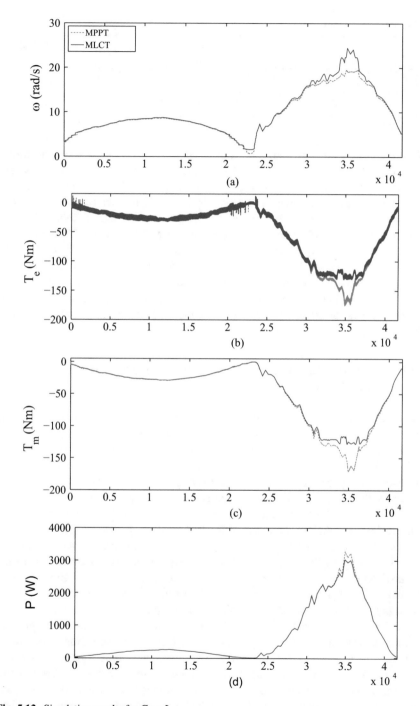

Fig. 5.12 Simulation results for Case I

Table 5.2 Results for Case I

	T_m		T_e		Energy
	Max(Nm)	D	Max(Nm)	D	(kWh)
MPPT	171.14	4.07×10^{-4}	188.54	1.2×10^{-3}	7.724
MLCT	127.96	1.15×10^{-5}	146.37	3.21×10^{-5}	7.606
%	74.77	2.82	77.63	2.58	98.47

power at the expense of high mechanical torque T_m and electromagnetic torque T_e. MLCT, however, greatly reduces the torque as shown in the subplots (a) and (b), with small difference in captured power compares to MPPT shown in subplot (c). Subsequently, the mechanical stress in the shaft is reduced, which leads to significant life extension and overall cost reduction through MLCT. Table 5.2 shows results based on Eq. (5.9) and the Miner's rule in (5.10). The MLCT strategy has 2.82% expected life expenditure of MPPT control, which means the life cycle of the tidal turbine blades will be extended to 35 times. Similarly, the life cycle of the shaft will be extended to 38.8 times with captured power a little less than MPPT during the life cycle of the tidal system with MPPT. Intuitively, because the life cycle is about 35 times longer, MLCT has the potential to harvest more energy than that from MPPT.

5.3.3.2 Case II: ANN-Based Sensor-Less Design and Speed Sensor Comparison

This case study compares the effect of using artificial neural network technique sensor-less design and that of speed sensor in estimating the tidal current speed. The simulation results are presented in Fig. 5.13, and the quantitative comparisons in Table 5.3 show the ANN and the speed sensor produce similar results. Although the result with speed sensor is smoother, this might be due to several reasons such as the model does not take into account the noise and delay in the sensor in real-time conditions and/or ANN needs to be trained further. However, the simulation results confirm the use of ANN networks to estimate the tidal current speed. The advantages of using ANN over sensor speed are as follows: (1) tidal current speed estimation is made directly by the ANN, (2) ANN only requires simple measurements of the captured power and rotating speed of the rotor to estimate the tidal current speed, and thus (3) ANN reduces the cost of installing and maintaining underwater sensors and significantly improves the reliability of the system.

Normally it is desirable for a tidal turbine generator to produce a ripple-free electromagnetic torque. These ripples could lead to speed oscillations resulting in vibrations in the shaft and shear stress. To reduce the torque ripples, a low-pass filter should be added at the input of the q-axis current in the machine control side, since q-axis current directly tracks the electromagnetic torque. Also, the space vector pulse width modulator (SVPWM) which subdivides the switching period into several states helps to synthesize the voltage vectors in order to generate minimum torque ripples.

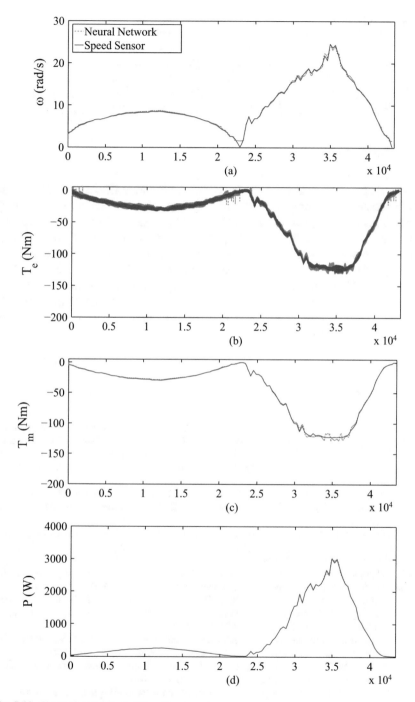

Fig. 5.13 Simulation results for Case II

Table 5.3 Results for Case II

	T_m		T_e		Energy
	Max	D	Max	D	(kWh)
NN	127.7	2.3×10^{-7}	131.0	2.3×10^{-6}	7.6
Sensor	122.8	1.1×10^{-7}	128.3	1.6×10^{-6}	7.6
Sensor/NN	96.2%	49.6%	97.9%	70.4%	99.95%

5.4 Summary

In this chapter, two control methods, MPEC for Smart-WEC and MLCT for tidal energy converter, are presented. The performance of the MPEC is verified through simulations under regular and irregular wave conditions. In the control strategy, the reference current is shifted to be in phase with the induced EMF for maximum electrical power transfer, without requiring sense coils for EMF measurement. Thereby, the efficiency of the device is improved. Subsequently, in the MLCT strategy, MPPT control is applied to the tidal system when incoming tidal speed range is low and a constant torque control when it's high. With this technique, the results show that, with small power trade-off, the life cycle of the system can be significantly extended.

References

1. H. Mendonca, S. Martinez, A resistance emulation approach to optimize the wave energy harvesting for a direct drive point absorber. IEEE Trans. Sustainable Energy 7(1), 3–11 (2016)
2. X. Xiao, X. Huang, Q. Kang, A hill-climbing-method-based maximum-power-point-tracking strategy for direct-drive wave energy converters. IEEE Trans. Ind. Electron. 63(1), 257–267 (2016)
3. J.S. Park, B.G. Gu, J.R. Kim, I.H. Cho, I. Jeong, J. Lee, Active phase control for maximum power point tracking of a linear wave generator. IEEE Trans. Power Electron. 32(10), 7651–7662 (2017)
4. P. Ricci, J. Lopez, M. Santos, P. Ruiz-Minguela, J.L. Villate, F. Salcedo, A.F.d. Falcao, Control strategies for a wave energy converter connected to a hydraulic power take-off. IET Renew. Power Gener. 5(3), 234–244 (2011)
5. V.J. Antonio, A.D. Montoya, G.S. Agustín, Increasing the efficiency of the passive loading strategy for wave energy conversion. J. Renew. Sustainable Energy 5(5), 053132 (2013)
6. D.E.A.M. Andrade, A. de la Villa Jaén, A.G. Santana, Improvements in the reactive control and latching control strategies under maximum excursion constraints using short-time forecast. IEEE Trans. Sustainable Energy 7(1), 427–435 (2016)
7. Z. Feng, E.C. Kerrigan, Latchingdeclutching control of wave energy converters using derivative-free optimization. IEEE Trans. Sustainable Energy 6(3), 773–780 (2015)
8. F. Fusco, J.V. Ringwood, A simple and effective real-time controller for wave energy converters. IEEE Trans. Sustainable Energy 4(1), 21–30 (2015)
9. N. Tom, R.W. Yeung, Experimental confirmation of nonlinear-model-predictive control applied offline to a permanent magnet linear generator for ocean-wave energy conversion. IEEE J. Ocean. Eng. 41(2), 281–295 (2016)

10. M.C. Sousounis, J.K.H. Shek, M.A. Mueller, Modelling, control and frequency domain analysis of a tidal current conversion system with onshore converters. IET Renew. Power Gener. **10**(2), 158–165 (2016)
11. A. de la Villa-Jaén, D.E. Montoya-Andrade, A. García-Santana, Control strategies for point absorbers considering linear generator copper losses and maximum excursion constraints. IEEE Trans. Sustainable Energy **9**(1), 433–442 (2018)
12. L. Ran, M.A. Mueller, C. Ng, P.J. Tavner, H. Zhao, N.J. Baker, S. Mcdonald, P. Mckeever, Power conversion and control for a linear direct drive permanent magnet generator for wave energy. IET Renew. Power Gener. **5**(1), 1–9 (2011)
13. P. Brooking, M. Mueller, Power conditioning of the output from a linear vernier hybrid permanent magnet generator for use in direct drive wave energy converters. IEEE Proc. Gener. Transm. Distrib. **152**, 673–681 (2005)
14. P.C.J. Clifton, R.A. McMahon, H.-P. Kelly, Design and commissioning of a 30 kw direct drive wave generator, in *IET Conference on Power Electronics, Machines and Drives*, Brighton, UK, Apr. 2010
15. M. Preindl, E. Schaltz, Sensorless model predictive direct current control using novel second-order PLL observer for PMSM drive systems. IEEE Trans. Ind. Electron. **58**(9), 4087–4095 (2011)
16. R. Vermaak, M.J. Kamper, Experimental evaluation and predictive control of an air-cored linear generator for direct-drive wave energy converters. IEEE Trans. Ind. Appl. **48**(6), 1817–1826 (2012)
17. T. Orekan, Z. Zhao, P. Zhang, J. Zhang, S. Zhou, J. Cui, Maximum lifecycle tracking for tidal energy generation system. Electr. Power Compon. Syst. **43**, 8–10 (2015)
18. M. Jackson, S. Umans, R. Dunlop, S. Horowitz, A. Parikh, Turbine-generator shaft torques and fatigue: Part I - simulation methods and fatigue analysis. IEEE Trans. Power Apparatus Syst. **98**(6), 2299–2307 (1979)
19. A. Secil, Fatigue life calculation by rainflow cycle counting method, Master's thesis, Middle East Technical University, Ankara, Turkey, 2004
20. A. Mullane, G. Bryans, M. O'Malley, Kinetic energy and frequency response comparison for renewable generation systems, in *2005 International Conference on Future Power Systems* (2005)
21. S.B. Elghali, R. Balme, K.L. Saux, M. Benbouzid, J. Charpentier, F. Hauville, A simulation model for the evaluation of the electrical power potential harnessed by a marine current turbine. IEEE J. Ocean. Eng. **32**(4), 786–797 (2008)
22. M.H. Beale, M.T. Hagan, H.B. Demuth, Tech. Rep., 2014. Available: http://www.mathworks.com/help/pdf_doc/nnet/nnet_ug.pdf
23. S. Bifaretti, P. Zanchetta, F. Iov, J. Clare, Predictive current control of a 7-level ac-dc back-to-back converter for universal and flexible power management system, in *Power Electronics and Motion Control Conference, 2008. EPE-PEMC 2008. 13th* (2008)
24. L. Kan, Z. Zhu, Online estimation of the rotor flux linkage and voltage-source inverter nonlinearity in permanent magnet synchronous machine drives. IEEE Trans. Power Electron. **29**(1), 418–427 (2014)
25. J. Belanger, P. Venne, J. Paquin, The what, where and why of real-time simulation, in *in Planet RT* (2010)

Chapter 6
Future Research

6.1 Introduction

Along with the increasing development of WPT, the security of power transferred has progressively become a major issue and has attracted the attention of
researchers. Since the energy is transmitted through the open magnetic field, it is
possible for all surrounding receptors, authorized or unauthorized, to gain access
to the energy. Hence, protecting the transmitted energy in various WPT applications
becomes a very important research area, especially in UWPT application for various
Navy facilities where security is a top priority, e.g., underwater warfare AUV,
monitoring systems, and ocean surface vehicles. Furthermore, without a secured
energy scheme, energy can be easily drained from Smart-WEC via UWPT system
by unauthorized means, such as unstable, illegal unpaid, and malicious adversary
receptors. In this case, energy security in Smart-WEC also becomes a critically.

Besides, security of the data exchanged is another major concerned that has
recently received significant attention. For instance, a cyber-attack, successful
or not, on a UWPT device located in a Navy technology could cause national
security threats. Hence, a layer of defense in UWPT to acquire perfect secrecy
communications is required.

6.2 Energy Encryption for UWPT System

Energy encryption for WPT system was first proposed in [1, 2]. The unique idea
is to utilize the characteristics of frequency sensitivity in WPT to encrypt the
transmission channel. Specifically, the switching frequency is purposely adjusted
based on a predefined regulation (which is unpredictable for all surrounding
receptors) and packed with various frequencies. Without the knowledge of the

T. Orekan, P. Zhang, *Underwater Wireless Power Transfer*, SpringerBriefs in Energy,
https://doi.org/10.1007/978-3-030-02562-5_6

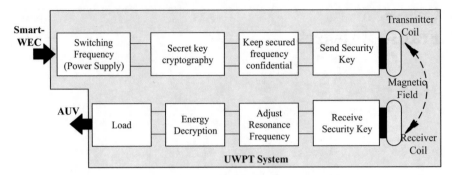

Fig. 6.1 Block diagram of the energy encryption scheme

secure sequence of session keys, an unauthorized receptor cannot receive power. Only when an authenticated device has a security key can power be transferred.

Applying energy encryption technique to UWPT system will significantly improve its applicability to security sensitive distributed ocean systems, such as Navy warfare vehicles. The block diagram of the energy encryption concept for UWPT is shown in Fig. 6.1.

At the initial stage, a secret key cryptography method is adopted to generate the security key for the energy encryption. With the generated key, the switching frequency is regulated and mixes with several frequencies, in order to keep the resonance frequency confidential and unpredictable from an unauthorized vehicle. If the device has authorization, charging request is accepted and the security key is sent. Then the resonant frequency is adjusted on the receptor using the received security key and matched to the frequency on the transmitter side.

Nevertheless, energy encryption technique for UWPT is still a challenging open problem, in the sense that, at the initial stage, there is no wireless communication between the transmitter and receptor (authorized or unauthorized), i.e., energy is received before the process of authentication begins. This leaves the UWPT system vulnerable to a malicious adversary receptor. One potential solution is to transfer data inductively between the transmitter and the receptor before energy transfer is initiated.

Meanwhile, for UWPT system with authorization, MPET control can be adopted to adjust the resonance frequency to match with the selected frequency value.

6.3 Underwater Inductive Wireless Data Transfer

Transferring power and data in underwater, at high data rate, while still maintaining high-power efficiency transfer will be the next hot research topic. Besides power, in many undersea application, there is a requirement to transfer data to and from the device wirelessly. Like inductive UWPT, inductive wireless data transfer is also based on electromagnetic field. The difference is when transferring data, signal-to-noise ratio instead of transmit-to-receive ratio becomes the most important factor. Usually, the latter one is more difficult to achieve than the former over long distance. Based on inductive data transfer, data signals are transmitted with the assistance of their mutual inductance. The signal transfer from the primary side to the secondary side of the UWPT is triggered by tuning a high-frequency magnetic field on and off, which corresponds to an amplitude modulation. In the inductive data transfer method, interference between power and data transmission must be avoided, and the internal voltage supply must be kept stable.

As shown is Fig. 6.2, data and power can be transferred simultaneously via a common inductive link. The technique was first proposed in [3]. The main idea is that an inductive coil transfers AC power through the transmitter and to receiver. In addition to energy being transfer, message signals (carrier wave) are being fed into the transmitting coil as well. The receiving coil reads the addition of the two waves and then distinguishes between them to read the data. One major advantage of using this technique in UWPT system is that conventional encryption method, commonly used in communication network, can be implemented. For instance, transmitter and receptor is assigned a private key pair beforehand, which then be verified through inductive data communication between the transmitter and receiver

Notwithstanding, in UWPT system, this method could be very challenging as carrier frequency cannot be higher than power frequency, due to the underwater environment. A feasible solution is to directly modulate the data via frequency shifting key (FSK) method and demodulate it on the receptor.

6.4   UWPT System Transmission Distance

Another inevitable research topic is how to transmit energy over a long distance in underwater. In the near future, the transmission distance between Smart-WEC and AUV or between multiple AUVs will be the most important technical challenge for researchers. This is a difficult problem to solve as the attenuation increases very fast when the distance between the coils increases. Design and optimization of coupling mechanism, and controlling radiation resistance of the UWPT system coil, will be a key solution to the problem.

6.5   Summary

The feasibility of energy encryption technique, and inductive wireless data transfer for UWPT system, is discussed in this chapter. The main idea of energy encryption is to generate a security key and then chaotically regulate the switching frequency of the transmitter power supply, such that only authorized receptor with the knowledge

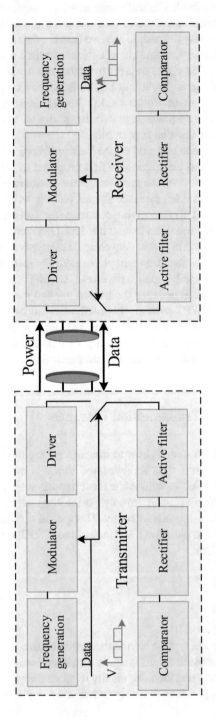

Fig. 6.2 Block diagram of wireless power and data transfer system

of the secured sequence can gain access. Meanwhile, inductive wireless power and data transfer via a common link is another method that can be adopted for UWPT system.

References

1. Z. Zhang, K.T. Chau, C. Qiu, C. Liu, Energy encryption for wireless power transfer. IEEE Trans. Power Electron. **30**(9), 5237–5246 (2015)
2. Z. Zhang, K.T. Chau, C. Liu, C. Qiu, Energy-security-based contactless battery charging system for roadway-powered electric vehicles, in *2015 IEEE PELS Workshop on Emerging Technologies: Wireless Power (2015 WoW)* (2015), pp. 1–6
3. J. Wu, C. Zhao, Z. Lin, J. Du, Y. Hu, X. He, Wireless power and data transfer via a common inductive link using frequency division multiplexing. IEEE Trans. Ind. Electron. **62**(12), 7810–7820 (2015)

Appendix A
Detailed Model of the Adaptive Controller

A.1 State-Space Equation

$$\frac{d}{dt}x(t) = \mathbf{A}_i x(t) + \mathbf{B}_i u(t)$$
$$y(t) = \mathbf{C}_i x(t) + \mathbf{E}_i u(t) \tag{A.1}$$

$$\begin{bmatrix} pv_{dc}(t) \\ pi_L(t) \end{bmatrix} = \begin{bmatrix} 0 & -\frac{1}{C} \\ \frac{1}{L} & -\frac{R_b}{L} \end{bmatrix} \begin{bmatrix} v_{dc}(t) \\ i_L(t) \end{bmatrix} + \begin{bmatrix} \frac{1}{C} & 0 \\ 0 & -\frac{1}{L} \end{bmatrix} \begin{bmatrix} i_{dc}(t) \\ V_b \end{bmatrix} \tag{A.2}$$

$$\begin{bmatrix} pv_{dc}(t) \\ pi_L(t) \end{bmatrix} = \begin{bmatrix} 0 & 0 \\ 0 & -\frac{R_b}{L} \end{bmatrix} \begin{bmatrix} v_{dc}(t) \\ i_L(t) \end{bmatrix} + \begin{bmatrix} \frac{1}{C} & 0 \\ 0 & -\frac{1}{L} \end{bmatrix} \begin{bmatrix} i_{dc}(t) \\ V_b \end{bmatrix} \tag{A.3}$$

$$\begin{bmatrix} v_{dc}(t) \end{bmatrix} = \begin{bmatrix} 1 & 0 \end{bmatrix} \begin{bmatrix} v_{dc}(t) \\ i_{dc}(t) \end{bmatrix} \tag{A.4}$$

$$0 = \mathbf{A}_{av} X + \mathbf{B}_{av} U$$
$$\mathbf{Y} = \mathbf{C}_{av} X + \mathbf{E}_{av} U \tag{A.5}$$

© The Author(s), under exclusive licence to Springer Nature Switzerland AG 2019
T. Orekan, P. Zhang, *Underwater Wireless Power Transfer*, SpringerBriefs in Energy,
https://doi.org/10.1007/978-3-030-02562-5

$$\mathbf{A}_{av} = D\mathbf{A}_1 + (1 - D)\mathbf{A}_2$$
$$\mathbf{B}_{av} = D\mathbf{B}_1 + (1 - D)\mathbf{B}_2$$
$$\mathbf{C}_{av} = D\mathbf{C}_1 + (1 - D)\mathbf{C}_2 \tag{A.6}$$
$$\mathbf{E}_{av} = D\mathbf{E}_1 + (1 - D)\mathbf{E}_2$$

$$\begin{bmatrix} 0 \\ 0 \end{bmatrix} = \begin{bmatrix} 0 & -\frac{D}{C} \\ \frac{D}{L} & -\frac{R_b}{L} \end{bmatrix} \begin{bmatrix} V_{dc} \\ I_L \end{bmatrix} + \begin{bmatrix} \frac{1}{C} & 0 \\ 0 & -\frac{1}{L} \end{bmatrix} \begin{bmatrix} I_{dc} \\ V_b \end{bmatrix} \tag{A.7}$$

$$V_{dc} = \frac{ED - RI_{dc}}{D^2} \tag{A.8}$$

$$I_L = \frac{I_{dc}}{D} \tag{A.9}$$

where I_{dc}, I_L, V_{dc}, and D are steady-state values of i_{dc}, i_L, v_{dc}, and d, respectively.

Using Eqs. (A.4) and (A.6), I_{dc} is calculated as

$$I_2 = \frac{\omega_0 k \sqrt{L_1 L_2} V_1 - R_1 V_2}{R_1 R_2 + (\omega_0 k)^2 L_1 L_2} \tag{A.10}$$

V1, V2, and I2 are expressed as follows by Fourier series expansion

$$i_{dc} = \frac{8}{\pi^2} \frac{\omega_0 k \sqrt{L_1 L_2} V_1 - R_1 V_{dc}}{R_1 R_2 + (\omega_0 k)^2 L_1 L_2} \tag{A.11}$$

A.2 Adaptive PI Controller

From Eq. (A.8), D is given as

$$D = \frac{E \pm \sqrt{E^2 - 4RV_{dc}I_{dc}}}{2V_{dc}} \tag{A.12}$$

$$D^* = \frac{E + \sqrt{E^2 - 4RV_{dc}^* I_{dc}}}{2V_{dc}^*}, \qquad 0 \le D^* \le 1 \tag{A.13}$$

$$C_{PI}(s) = K_p + \frac{K_I}{s} \tag{A.14}$$

Appendix B
Tidal Generator Detailed Model and Parameters

This appendix provides the detailed model and parameters of the tidal turbine system.

B.1 Detailed Model of the Tidal System

The turbine model is

$$P = 0.5\rho\pi R^2 v^3 C_p(\lambda) \tag{B.1}$$

$$C_p(\lambda) = c_5\lambda^5 + c_4\lambda^4 + c_3\lambda^3 + c_2\lambda^2 + c_1\lambda + c_0 \tag{B.2}$$

where c_0 to c_5 are constants, ρ is the water density, and C_p is the tidal turbine power coefficient. λ is the tip speed ratio, which is defined by

$$\lambda = \frac{\omega R}{v} \tag{B.3}$$

The model of the permanent magnet synchronous generator is

$$v_{ds} = R_s i_{ds} + L_d \frac{di_{ds}}{dt} - p\omega L_q i_{qs} \tag{B.4}$$

$$v_{qs} = R_s i_{qs} + L_q \frac{di_{qs}}{dt} + p\omega L_d i_{ds} \tag{B.5}$$

$$P_e = \frac{3}{2}(v_{ds} i_{ds} + v_{qs} i_{qs}) \tag{B.6}$$

© The Author(s), under exclusive licence to Springer Nature Switzerland AG 2019
T. Orekan, P. Zhang, *Underwater Wireless Power Transfer*, SpringerBriefs in Energy,
https://doi.org/10.1007/978-3-030-02562-5

$$T_e = \frac{3}{2} p \psi_m i_{qs} \qquad (B.7)$$

where v_{ds} and v_{qs} are the q-axis and d-axis stator terminal voltage, respectively; i_{ds} and i_{qs} are the q-axis and d-axis stator currents, respectively; L_q and L_d are the q-axis and d-axis inductances of the PMSG, respectively; R_s is the resistance of the stator windings; ω is the electrical angular velocity of the rotor; and p is the number of pole pairs of the PMSG.

The model of drivetrain is

$$J \frac{d\omega}{dt} = T_m - T_e \qquad (B.8)$$

$$T_m = \frac{P}{\omega} \qquad (B.9)$$

The model of machine side converter with controller is

$$i_{ds}^* = 0 \qquad (B.10)$$

$$\frac{dm_1}{dt} = i_{ds}^* - i_{ds} \qquad (B.11)$$

$$v_{ds}^* = k_{P1}(i_{ds}^* - i_{ds}) + k_{I1}m_1 - p\omega L_s i_{qs} + R_s i_{ds} \qquad (B.12)$$

$$\frac{dm_2}{dt} = \omega^* - \omega \qquad (B.13)$$

$$\omega^* = g(v) \qquad (B.14)$$

$$i_{qs}^* = k_{P2}(\omega^* - \omega) + k_{I2}m_2 \qquad (B.15)$$

$$\frac{dm_3}{dt} = i_{qs}^* - i_{qs} \qquad (B.16)$$

$$v_{qs}^* = k_{P3}(i_{qs}^* - i_{qs}) + k_{I3}m_3 + R_s i_{qs} + p\omega L_s i_{ds} + p\omega \psi_m \qquad (B.17)$$

where k_P and k_I are the proportional and integral gains of the PI controller and ψ_m is the flux linkage generated by the permanent magnet.

The model of the filter is

$$u_{dg} = v_{dg} + L_g \frac{di_{dg}}{dt} - \omega_g L_g i_{qg} \qquad (B.18)$$

$$u_{qg} = v_{qg} + L_g \frac{di_{qg}}{dt} + \omega_g L_g i_{dg} \qquad (B.19)$$

The model of the grid side converter with controller is

$$i_{qg}^* = 0 \tag{B.20}$$

$$\frac{dm_4}{dt} = i_{qg}^* - i_{qg} \tag{B.21}$$

$$u_{qg}^* = k_{P4}(i_{qg}^* - i_{qg}) + k_{14}m_4 + \omega L_g i_{dg} + v_{qg} \tag{B.22}$$

$$\frac{dm_5}{dt} = v_{DC}^* - v_{DC} \tag{B.23}$$

$$i_{qg}^* = k_{P5}(v_{DC}^* - v_{DC}) + k_{15}m_5 \tag{B.24}$$

$$\frac{dm_6}{dt} = i_{dg}^* - i_{dg} \tag{B.25}$$

$$u_{dg}^* = k_{P6}(i_{dg}^* - i_{dg}) + k_{16}m_6 - \omega L_g i_{qg} + v_{dg} \tag{B.26}$$

The model of DC link is

$$C\frac{dv_{DC}}{dt} = \frac{P_t}{v_{DC}} - \frac{P_g}{v_{DC}} \tag{B.27}$$

B.2   Torque-Time Curve and Rainflow Counting Method

Figure B.1 shows a typical torque-time curve, which may occur for a given shaft. Each identifiable positive torque peak (labeled A, B, C, D, E) is counted as one cycle of a torque equal to the value of that peak. The rotation of Fig. B.1 by 90° forms Fig. B.2, which shows the half-circles and its magnitude. In this figure, the raindrop goes from the starting point S through P_1 and ends at the V_2 dashed line because V_2 is lower than the S. The same mechanism works for the raindrops from V_1, V_2, V_3, etc. The raindrop from the peak P_1 goes through V_1, V_2, and V_3 and ends at the P_4 dashed line because P_4 is higher than the starting peak P_1. The raindrops from P_2, P_3, and P_5 stop at the paths of a previous raindrop. Hence, the difference between the starting point and the ending point is twice the magnitude of that half-cycle. Note that this is only a statistical estimation caused by stress. In reality, other factors such as corrosion, which is out of the scope of this dissertation, could also result in a shorter life.

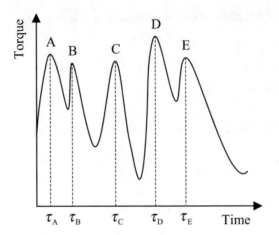

Fig. B.1 Torque-time curve

Fig. B.2 Rainflow counting method

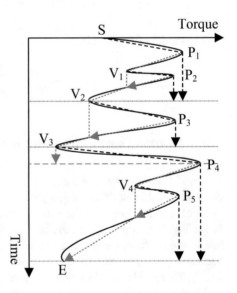

B.3   Artificial Neural Network

Multiple neurons in a hidden layer of an ANN with a vector input are shown in Fig. B.3, where $p_1, p_2, \ldots, p_R$ are elements of the input vector $\mathbf{P}$ and the weight vector is $\mathbf{W}$.

Fig. B.3 Multiple neurons in
one hidden layer

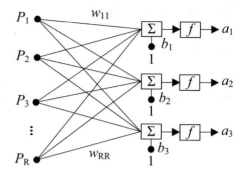

The output **a** is obtained by

$$\mathbf{a} = \mathbf{f}(\mathbf{Wp} + \mathbf{b}), \tag{B.28}$$

$$\mathbf{a} = \begin{bmatrix} a_1 \\ a_2 \\ \vdots \\ a_R \end{bmatrix}, \ \mathbf{b} = \begin{bmatrix} b_1 \\ b_2 \\ \vdots \\ b_R \end{bmatrix} \tag{B.29}$$

$$\mathbf{W} = \begin{bmatrix} w_{11} & w_{12} & \cdots & w_{1R} \\ w_{21} & w_{22} & \cdots & w_{2R} \\ \vdots & \vdots & \ddots & \vdots \\ w_{R1} & w_{R2} & \cdots & w_{RR} \end{bmatrix} \tag{B.30}$$

where **b** a scalar bias and **W** are adjustable so that the neural network is trained
accordingly. These parameters are trained using backpropagation method or extreme
learning machine for fast-speed applications. f represents the transfer function used
in the layers for different purposes.

B.4   Tidal Data Table

See Table B.1.

Table B.1 Data for reference
generation stage

Tidal current speed (m/s)	MPPT rotor speed (rad/s)	MLCT rotor speed (rad/s)
0.1	0.72	0.72
0.2	1.44	1.44
0.3	2.16	2.16
0.4	2.88	2.88
0.5	3.60	3.60
0.6	4.32	4.32
0.7	5.04	5.04
0.8	5.76	5.76
0.9	6.48	6.48
1.0	7.20	7.20
1.1	7.92	7.92
1.2	8.64	8.64
1.3	9.36	9.36
1.4	10.08	10.08
1.5	10.80	10.80
1.6	11.52	11.52
1.7	12.24	12.24
1.8	12.96	12.96
1.9	13.68	13.68
2.0	14.40	14.40
2.1	15.12	15.12
2.2	15.84	15.84
2.3	16.56	16.62
2.4	17.28	18.72
2.5	18.00	20.73
2.6	18.72	22.69
2.7	19.44	24.62
2.8	20.16	26.52
2.9	20.88	28.41
3.0	21.60	30.28

Printed in the United States
By Bookmasters